May Beetle or Africa's in Siberia

Author: Victoria Korchikova-Malovichko,
Viva Vichka and Rocky Fokichrok

CONTENTS

1. Preamble.
2. Foreword.
3. Chapter Zero. The Winter Dream of Iya. Rendezvous.
4. Chapter One. The Summer Dream of the Trio. Convention.
5. Chapter Two. The Spring Dream of Victor and Iya. Africa in Siberia.
6. Chapter Three. The Autumn Dream of Victor and Iya. The Centennial of VOVAA-Vicinema.
7. Chapter Four. Off-Season Dream. The Structure of the Vicinema Academy.
8. Afterword. Iya's Letter and the Airlift

PREAMBLE

- Mom, look! Look how the obstacle has accelerated them! Look! Look how they hit, push off, and rise up!
- I see, son.
- But... look here now! At the others! And they slowed these ones down! They've even stopped them!
- Yes, for these, it's a hurdle. For others it is an uphill climb and advancement!

FOREWORD

To fish - water, to birds - air, to man - the whole earth, and to the superhuman - both water and air and the whole earth.

Iya was sitting on her favorite meadow in the woods at a maple tree, gazing sadly at the native ant house. She had long ago stopped being surprised by the behavior of the ant family, the large and shiny black ants that belong to the Lasius niger species. Reading Iya's thoughts, sensing her mood, they stopped gallantly today near her involuntarily upended hand, near the ant slide and, as best friends are supposed to, sensing her sadness and bitterness of experience, guarded her concentrated mood of sadness. "Nigers" hurried to inform others, nearby ant species of the important guest and her eager pacification. The telepathic work was done, as usual, to perfection. Iya lay down under a maple tree and in about twenty minutes, representatives of the four species of ants outlined all four of her seven aura layers around her.

- How could it be? - Iya's thoughts involuntarily burst

out in indignation. - Why such a rejection? After all, they are educated and polished people of science! How could it be? How is it so? Why so much ideology? So much politics in science? They used to say that there are more prospects and freedom over the ocean... How could it be?

How come…

Ugh! What's that slime?

Iya lifted her head sharply and shook the slug off her wrist, which had already actively begun its work. Suddenly she noticed how diligently four groups of ants were working, marking the borders of her aura.

- How can we ignore and reject the wonders around us? Why do these professors and experts limit themselves and our possibilities so much? Are they afraid for the people? They have not cared about them for a long time! They are interested in politics and finances! They utter evolution, but broadcast from their mental prisons. Just think: they wonder like schoolchildren, indignantly declaring: "It is contrary to the laws of physics! Do you understand that, Iya?! It's against the laws of gravity! Do you understand that, Iya?"

They talk to me like I'm crazy or like a schoolgirl who hasn't learned her lesson.

- Yes! Of course! It is against the laws of physics! But are those laws of physics true? Have they ever questioned or re-checked them?

They should also add: "The Church will burn you at the stake tomorrow if you make such a statement in public again! Or we will send you to an asylum if you once again doubt the truth of the laws of physics..."

Iya took a deep breath. She squinted against the bright light of the sun that had already risen over the maple tree and instantly immersed herself in enjoying the sensations and sounds of her home glade. The ants, like faithful guardians, continued to outline the wave emissions of Iya's aura, still diligently protecting her biofield from the intrusion of any other insects.

From the rhythmic and melodic sounds and vibrations heard only by Iya, she imperceptibly lifted herself off the ground in the distance of the grass that had grown, which now served as a thin downy mattress for Iya already dozing in the rays of the midday sun.

Feeling and registering the transition from excitement to peace and long-awaited tranquility, Iya was about to enjoy the minutes of soaring when suddenly her body slammed to the ground.

- What now?! - Iya opened her eyes and saw herself and an old, gray-haired male being leaning over her lips.

- Oh, yes! I forgot again! I haven't cleared my memory and I'm bored again. I wish I knew what he looked like; however, that won't be necessary. My "antennae" will recognize him when I see him... I just need to clear my memory of meetings with the professors here, for he wants to kiss me, and I was about to tell him about my slurs.

- Tell me what the insults and upsets are, - a man's voice suddenly sounded above Iya's head.

Iya jumped up abruptly and immediately disrupted the harmonious rows of all four species of ants. In a panic, they quickly mingled and began to scatter in their directions.

- Overheated. I forgot to put at least a leaf on top of my head again, - Iya whispered to herself in a frightened whisper, looking around.

She rose to her full height. Straightened up. She calmed down. Turned to face a maple tree. She leaned against it and, closing her eyes, stroking the trunk of the tree with her hands, found the source of maple sap. She gently pressed her lips to the already soaked bark and, enjoying the maple moisture, returned to her physical body. The moisture sprinkled her mouth and a thin stream of juice smothered her thirst. Her body relaxed with pleasure, her shoulders slumped. Iya threw her head back, raised her arms up, and spun around, lowering her arms as if to land. With another wave of her hands and a sigh, she crouched under the very trunk of a maple tree on its roots, continuing to reason aloud:

- Still, maybe they were right. I mean, there was a kind and calm professor there. Maybe he was right. After all, he was so kindly trying to tell me that the path I had chosen would ruin me. Maybe I really went in the wrong direction? Maybe this is really the road to the unknown and the social gulley! After all, I'm nobody now! And no one will believe me that people can fly or at least create antigravity machines. Probably, I should start with the simple and accessible for all. With something that has already been told a hundred million times, and retold, and approved, and confirmed, and defended... A dragon of some kind? Oh, a drone or a quadro-zadron. Oh, I'm getting really hot. Maybe I could just add a fifth element - a propeller of some kind - or remove it altogether, and it would be a quanto-drome! Oh! Exactly! Although it should just be the Pentagon! Whoa! That one's got to go! I'll call it the Pentagon, the Dragon Five-Winged. Everyone will be heading for love it. There's an element of magic, and now it's fashionable to look for meaning in fairy tales, and there's also the Pentagon. And it's gonna be great! Just give this insatiable breed one more chance to...

After all, it's hard for chubby and old people to think! Of course it's hard! They've got candida sitting there, begging for candy. And I'm telling them about flying!

The aggressively negative flow shook all of Iya. Under the weight of indignation, criticism, and denial, she slid like a caterpillar down the trunk of the maple down to the roots again. Taking a big breath, she suddenly felt the sweetness of the air. The sun had already warmed the resin on the trunks of the nearby pines, and dragonflies and May beetles were adding to the music of the forest with their percussion.

- Trshch, zhuuu, zzzzz, zzzzz..." - the members of the forest orchestra were buzzing and bustling.

Iya had already covered her eyes with pleasure, closing under the weight of her enormous terry black eyelashes, and through them she enjoyed the flecks of indistinct sunlight...

Suddenly something black jumped up right under her nose. Out of surprise, she cringed frightened into the trunk of a maple tree with her eyes wide open, and brought them down to the tip of her nose. Sharpening her gaze, she finally got a good look at the spider.

- Phew! - Iya sighed in relief, dilating her pupils into place. - Big, but not poisonous and not really scary. Although how disgustingly black it is. -

And after a short pause, she consciously sadly added: - Just like my thoughts.

The spider once again did a somersault, already at the level of the Iya's nose, then like a pendulum swung a little before her eyes and, having found a point of balance, as if with reproach and warning looked at Iya point-blank. Then he lifted his heel, emitted a liquid from it, and, connecting with the top, somewhere in the sky, waiting for the wind, prepared himself for free flight.

All the while, Iya pressed her back even tighter against the tree, as if wanting to move the maple, and without blinking, opening her eyes wider and wider, watching the black spider's every movement.

At last the right wind was blustered. The spider raised its heel once more, and soared upward with pride and determination, gladly surrendering to the freedom of movement.

Exhaling, Iya, flapping her eyes, lay down comfortably on the soft grass by the roots of a maple tree and said out loud in her heart:

- Even spiders can fly! Why, why do I only fly in my dreams?!

What's wrong with the dreaming? Why can't it be in the real world?

CHAPTER ZERO

The Winter Dream of Iya.

Rendez-vous

Moscow. Monument to Bulat Okudzhava.

It's winter. It is cold. Uninhabited pre-New Year's Arbat.
Evening.

The song "Wings" is played. The phrases of the melody trail behind Iya, aerial lightly over the pavement of the capital's central street. The evening mist from the winter lanterns envelops the silhouette of Iya, hovering joyously and freely between the artificial trees in the center of Novy Arbat, decorated with garlands of light bulbs...

- Hello, Professor, I'm already there.

- Yes, Iya, I'm late, I'm sorry. I'm looking for your present. Haven't seen you in ages. I'm at the bank. Find a cafe nearby, don't wait for me in the cold.

- Don't worry, Professor. I'll go to the bookshop on Arbat and buy your books. I came to get your autograph.

- I'm afraid I must take yours. I've heard of your fantasies.

- Don't joke like that, Professor.
- I'll see you soon. I look forward to seeing you. Go to the new coffee house, next to Okudzhava's monument, listen to his songs...

Almost inaudible is the song to Bulat Okudzhava's lyrics "The Musician Was Playing the Violin". Iya, slightly distracted by the words of the song: "I'm not that curious - I was flying through the sky", hurried to end the phone conversation with the professor.

- Don't worry, Professor, there's no hurry. I am even glad that you invited me to Arbat, I have a couple of hours to be alone with Moscow. And what's more, it's so lucky: it's completely forsaken. I'll choose a place where we can have some privacy.

- All right, enjoy the capital for now.

Cafe. A Soviet-style cafe. Iya settled in a corner at a table by the Christmas tree, leaning against the wall decorated with popular Soviet-era items: vinyl records, musical instruments. A flushed, clean-shaven and trim, but already gray-haired professor immediately found Iya in the jam room.

- Can I congratulate you? You defended your PhD's thesis? Did my recommendations do you good?! - As he was undressing, the professor hurried to hug his student.

- I'm afraid to disappoint you, Professor. I've been blown into something I don't even know what it is. But I'm sure it's something big and important.

- Are you ready to defend another doctoral thesis soon? Have you finally come to invite me as an opponent as well?

- It's all I had ever dreamed of, Professor. I couldn't have wished for greater happiness...

- But just what? Did something happen?

- Professor, at least promise me that you won't stop talking to me if I tell you everything.

- Have you made a discovery? Dear Iya! I knew, I knew you were worth the effort and the time. I immediately saw the great potential in you. It was not for nothing that I asked you to assist me and even replace me at the lectures! And what a success! And what a triumph! Don't be tightfisted! Well! Speak up! Have you found your way in quantum physics?

A waiter of Mongolian-Tatar appearance came up, quickly put the menu on the table and immediately asked unceremoniously:

- What do you want your "varenyk" with it?

Iya hassled and straightaway started making excuses:

- Excuse me, Professor, I know that you like shashlik, but while I was waiting for you, something instantaneously drew me to the Soviet atmosphere after the bookshop on Arbat. I saw things from my childhood and dumplings and varenyks in the window, and for some reason I remembered my grandmother and comfort...

- It's okay, it's okay. Just in time for Christmas to fast a little bit. And I'll dive back into my childhood and order some fish, too. I was born near the White Sea.

- Oh! Really? I spent my whole childhood on the Black one.

- So much for the thesis and antithesis, - muttered the professor, smiling and looking at the menu.

The waiter stood, irritated and showing with all his appearance that he had to hurry to another table.

The professor and Iya, smiling and looking at each other, said almost in a whisper in sync:

- Apparently, they took the worst from the good times.

- My dear... - The waiter looked down at the professor studying the menu, and after a pause, without raising his eyes, he continued: - The fish here is prepared with dignity?

- What? - the waiter, named Izmood, which was written on his badge, asked with a wry smile, hurrying and rushing to move to another table.

- What kind of fish do you serve? - The professor translated his own question, adding firmness to his voice.

- On the backside, - the waiter cut off dryly.

Iya sighed, smiling and trying to lighten the atmosphere, but she still quipped:

- And they serve the fish on the back side.

And, pouring herself a light brown tea that had been served an hour ago, offered it to the professor as well.

The professor glanced at the tea, raised his eyebrows, and grinned through his teeth:

- What an exact replica of Soviet poverty.

- But no sugar was spared. - Then Iya looked up at the waiter and said: - And here it's the other way around, like in Wonderland.

- What? - the waiter asked irritably and with an accent again.

- I asked for sugar-free and sea-buckthorn tea.

- I told you, we don't do that. It's the state standard and prepared according to the requirements of GOSTs.

The professor choked up with laughter and hurried to share his impression:

- I live nearby and didn't even realize there is a direct portal to the past under my windows.

- Will you be ordering fish? - And, without waiting for an answer, the waiter Izmood headed to another table, throwing over his shoulder as he went: - Call me, when you're ready.

Iya sighed, glanced at the clock on her two phones and immediately tried to hide her anxiety over the missed calls.

- Yes, it's a little late even for dinner. - The professor noticed Iya's apprehension and anxiety.

- No, not at all! - Iya hurried back to her calm frame of mind. - We haven't seen each other in so long, and I haven't told you the most important thing.

- The most important thing, as you remember, is to define the antithesis and everything will be in its place. - The professor, looking at the menu, lifted his eyelids and as if by the way spoke slyly: - And you are afraid to cross me and confess that you've become the opponent of my long-time published theory, aren't you?

- Yes. - Iya fussed again. - I'm sure it would be right to choose dumplings, but not of the same kind, but with different fillings, - she said quite seriously and also looked slyly at the professor.

- You won't pass GOST. - The professor easily supported Iya's game and added: "And fish is what the doctor prescribed for time and health".

- And who, who sets these GOSTs? - Iya lit up and immediately released all her emotions, hastening to add: "Well, let them make up the GOSTs - they have to do something. All right, let them make up GOSTs, they have their own job to do, but check them! Did they ever dare to check? Have they ever questioned the established and adopted norms and laws?

- That's why there are norms and laws, so that everyone should follow them. Not everyone is like us. People are different. So in order to keep them sane, there has to be order. Order in their language... And for that time, for those centuries, that was the plan. You couldn't reveal everything. But I'm glad you did justify my... - A surge of voltage and a flicker of light interrupted the Professor's quick explanation.

At that moment, the waiter grew out of the ground:
- Dear guests, sea buckthorn and sugar-free tea for you!

- And what about the GOSTs? - Iya opened her mouth and already looked at the professor in confusion.

In front of them stood a waiter of the same appearance, but with a completely different gut. Smiling and with a name on his badge: "Bezmood". He put down another cup for the professor and gallantly poured him sea buckthorn tea as well.
- I'll have a sea-buckthorn tea with you, Iya, - said the professor. - And then, turning to the waiter, resolutely asked: "Please, bring me that rainbow fish. And for the lady..."

- And for the lady, I will order different dumplings, also rainbow ones. - Quickly, in a hurry to complete the order, Iya turned to the waiter: "Can you change or correct the GOSTs for the guests?"

- As an exception, but payment according to GOSTs all the same. Sorry, or rather, according to the tariff. - Continuing being gallant, Bezmood took the menu and the snow-white tea set with the stamp on the tray: "Made in the USSR, GOST #9669" off the table.

Interrupting their conversation for a while, the professor and Iya, leaning away from the table, letting Bezmood clear the table, and at the same time noticing and appreciating the interior of the cafe with all the accessories of the Soviet and pre-Soviet era: from the samovars and accordions hanging on the walls, to the holdovers and Soviet phones on shelves of cabinets with a selection of books that formed the indispensable attribute of the Soviet intelligentsia. Memories of Soviet life, of constantly working women and men rushing around somewhere, came to mind. Then one could see queues, traffic jams, for some reason wide and open prams with waste paper and bottles, and newspapers everywhere: in the toilets, behind portraits, in window frames and under the legs of truffles for keeping things in balance... Newspapers were here too, also in the hands of old people who were suddenly crowding the best reserved seats in the pub by 9 pm along with a group of noisy younger people who happened to pop in here...

Iya, being still under the influence of the Soviet era and in the image of a working woman, quickly blurted out:

- And the bill then, and the bill is on me. - She glanced happily at the professor and, being proud of herself, straightened up and announced majestically: - Dear Professor, please allow me to pay for our order!

- Iya, breathe out. The masculine and the feminine have not yet been abolished. Don't reject your own nature. You'll lose your temper, you won't get to the bottom of it. And don't waste time, get out your papers, they're all crumpled up from your hesitation.

While the fish is being cooked, let's take advantage of the space and encircle ourselves in silence for a moment, distance ourselves from the audience that has suddenly appeared around us and get down to business.

She recognized the kind old professor again, who always understood what Iya wanted. But she was still discouraged and could never get used to the fact that there was another person like her in the world - one who could see. Admiring the professor's telepathic abilities and directness, she pulled out all her intentions without a word.

- Here is the anti-gravity platform. - Whitening from the fear and cold, Iya took a deep breath.

- Having recognized and found my books, and even under a pseudonym?! - The professor, quickly unfolded Iya's drawings, studied the calculations, while simultaneously commenting on what he saw, smiling with the satisfaction.

- How could I not know! You were the only one at the meeting who supported entomologist Victor Grebennikov, while all his friends-physicists denied his discovery and even invention.

- How do you know what I refuted? - Without raising his head and without taking his eyes off the drawings, the professor was making some notes in his notebook.

- But, in general, they wrote that they did not recognize him, - answered Iya, scared stiff.

- Different variations of anti-gravity platform models have already been created today. And experiments continue. But it is too early to announce it even now, in the twenty-first century. Both then and now is not the time.

- Is that why you were so critical of me?

- Not you. Not you either. Realize your feminine essence and purpose first, and then perform the mission of super-…an. Otherwise, you are "It" now.

- What? "AN"? Oh... UN! I mean, "IT"?

- I love your humor, but I think we're already getting rainbow fish and your colorful "vareniks".

- Oh, really! Look, they are really colored ones. - Iya, fixing her hair, looked with interest at the tray, that Bezmood had just brought to the table.

- Nice, right? The green ones are spinach, but the red ones? What are they with? - The professor was looking at Iya's order with curiosity.

- Mmm, amaranth or beetroot? How interesting. - I was impatient to taste the dish.

- It's interesting that you're a real Russian.

- Why is that? - Iya stopped devouring her food.

- Only a true Russian today would call "a beet" as "a buryak" instead of "svekla".

- Do you think it's because of solidarity? - Iya, cleverly concealing her opinion, answered the question about politics with the same question.

- I don't know. All I know is that everything is moving and stirring, like in your plate of colored 'vareniks'. What do you think of Moscow today? How long have you been living across the ocean?

- During eleven years, Moscow has turned into my plate of colored 'vareniks'. On the surface and on the first side, it looks smooth and clean, but inside... And inside there are a lot of colors and some incomprehensible elements... - Iya, without raising her head, devoured the 'vareniks' with the appetite of a gourmet.

- It's fried onion with dill and something else...

- Yeah, that's a lot of spice in one plate.

- To keep your nose in the wind and not fall asleep on the move...

- We've been distracted, - Iya muttered, chewing.

- And what do you think, Iya, they will accept the idea of flying?

- I think if you give them the command and order to fly, they will fly.

- Ah! Yes, that's right, you're right, Iya, without a second thought, they'll fly.

- And with those who still think and are afraid, - Iya took her eyes off the food and went on asking the professor, - with those who are full of fears, complexes, rules, who have been heavily cultivated by socialization in kindergartens, schools, universities, factories and ministries, - what to do with them, professor?

- With all those who are created according to GOSTs?

The waiter appeared again as if from under the ground:

- Are our guests happy? - leaning in slightly, he asked, smiling.

- Yes, we are ready to thank you soon and leave, - said the professor without pauses. He leisurely continued to cut the fish and, without raising his head, with a slight movement of his shoulder energetically cut off the waiter.

Iya, picking out the colors of the vareniks and being serious and focused preparing to finish the conversation, similarly sent the waiter away:

- We'll call you when we need something.

- What happened to you? Where on earth did you get lost? - The professor paused in his work on the fish and looked into Ia's eyes.

- In my dream, Professor.

- And seriously?

- Dreams, to be exact.

- Okay, weed is good, - the professor retorted, pulling the rosemary from the belly of the fish. - Have you become a true Canadian? But you haven't lost yourself! I know you haven't!

- Have you ever flown in your dreams? Do you remember? - Iya continued to question.

- Yeah, I must have flown when I was a kid.

- No, for real.

- In your dreams and for real? Do you drink or smoke? Or are you relaxing today?

- I'm quite serious, Professor. I'm flying. Do you understand?

- Flying in the dreams for real?!

- Believe me! It's an indescribable feeling of freedom and life. Full life. Fulfillment life!

- Okay. Okay. Without the state standards? Without the GOSTs?

- Almost.

- So, free in the family and outside of society. How did you manage to do that: in the woods? In nature? Managed to escape... from socialization?

- Managed to come running back to myself. I'd rather say that.

- Disappointed now by a member of the Academy of Sciences, you are my dear student, but charmed at once by me, a professor of physics and philosophy.

- Is it bad now? You were my last hope!

- Listen carefully! That's why I'm inviting you to a scientific symposium with me. Without you, I wouldn't have dared such madness. But if you, on your own, have proved what I have been working on for so long, then who, if not you, deserves to assist me in the flight?

Iya has burst into tears.

- Sorry, emotions.

- Well, well, well. Now I see they didn't send a robot from America, you're still alive. Well done! I look forward to seeing you at the symposium. That's it! Go, it's late, or you'll go completely 'soft'. Straight on out. Our friend-driver has been waiting for you for a long to bring you back. It's cold now, and you've gone 'soft'. You won't take off. Do not fall apart!

- Thank you, Professor! I adore you! What about the autograph?

- After the flight. And choose only one book.

- Why just one? - Wiping her eyes and sighing deeply like
a child, Iya interjected, relieved of her tension and frustra-
tion.

- When you read them, you understand it. This is my gift
to you.

- Thanks! Hence, I choose Africa in Siberia.

- Read it. All after the flight, May Be-etle.

- And, I choose 'May Be-etle'.

CHAPTER ONE
The Summer Dream of the Trio.
Convention.

Scientific international symposium.

Banner: "Against the Laws of Physics".

Sunny day. Serenity. In a foyer with a panoramic view of the sea there are three persons.

Iya, standing at the huge window, looks out, smiling, into the distance, sensing someone forthcoming from behind.

Two men are in the background in far different corners of the foyer.

Feeling his oncoming with her spinal cord, Iya counts to three with a smile, telling herself in a whisper:

- One, two, three, go.

At the same moment Victor approached Iya as if on signal.

And his inner voice rang out: "Hello, here we are again."

Iya, just as well as Victor, without opening her mouth and without moving her lips, looking into his eyes with a slight smile, replied: "Hello, how nice that in this life you are He and I am She, there is a chance to correct the mistakes of the past incarnations".

Suddenly, after a quick look around, they looked over at each other, a little stunned by the sensation of another stream of information nearby. And in one inner voice without voice or facial expression they said together just as telepathically: "We need to act lucidly now... Someone is watching".

Victor: I'm home, - he smiled and stretched ironically, glancing at Iya.

Iya: And you, too?

Victor: Local?

Iya: Is it?

Victor: When you say no, but your figure and casual manners say yes.

Iya: Wrong.

Victor: But I know for a fact that you can fly.

Iya: I've always dreamed about it and now I'm flying.

Victor: Do you have your wings with you?

Iya: Hiding.

Victor: Why?

Iya: Not yet. They won't understand. But I always carry it with me.

Victor: And you, too? Is this yours? - pointing to a suitcase in which he guessed the same flying machine as his...

Iya: Yes. Why "you too"?

Victor: I like to play around, too. It's not scary. It's easy to float over the sea.

Iya: And where do you usually hover? - Not surprised by Victor's clairvoyant and telepathic abilities, Iya continued the dialogue.

Victor: Over "Africa in Siberia". Above its vastness. - Just as unsurprised by Iya's questions, Victor continued to communicate with her as if they had already lived together all their lives.

Iya, without waiting for the last phrase, looking at the expression on Victor's face, already burst into laughter.

Iya: You're also funny.

Victor: And before the funny what? Why else?

Iya: I wanted to say something nice, but I remembered it was too soon.

Victor: It's no use playing these games with me.

Iya: Which "these"?

Victor: Man-woman, mating dance, foreplay, trap-and-trap...

Iya: It was too brazen now.

Victor: I know. I was offended and even smelled it right away. So you gave yourself away.

Iya: Are you reading my mind, too? - Iya dared to ask again, just in case.

Victor: And how do you know that?

For a second, the strong attraction made Victor and Iya

freeze, and everything around them seemed to come to a halt. They involuntarily reached out to each other, but the magnetism was redirected in time to the third.

Professor Carsten von Münhausen-Dach: Oooh, friends! Now everyone will be back from breakfast and they can intercept your happiness at once, Miss Frau and Mr. Herr. Is that how your mother and father were called at court in Russia? Isn't that right?

Iya: Ha-ha! Whoa!

Victor: Do I know you?

Carsten von Munchausen-Dach: We'll be friends soon, but for now, okay, let's make out a little.

Iya: Do you also know everything?

Carsten von Munchausen-Dach: Sorry for peeking, but the glow and the attraction were so noticeable that I only hastened to warn you that if you continue like this, you will immediately get lit up. And we have a lot of things to do together!

Iya: OK, Professor, do you fly too?

Professor: Not as much as you, but we'll still have time to discuss it. Come into the hall. I have to go. It's my speech tonight.

Iya: Wait, you mean telepathy after all?!

Carsten von Münchhausen-Dach: And many more different tricks.

Smiling enigmatically, the professor took a small sketchbook out of his suitcase and pulled it apart with a slight movement. He ascended the platform and glided into the conference room, passing the amphitheater and landing behind the pulpit.

Victor and Iya, having done the same with their antigravity platforms resembling exactly the same sketchbooks, followed the professor without surprise but happily, looking around to see if anyone had noticed them in the corridor.

Carsten von Münchhausen-Dach: It's wonderful that we've already found it. The first task is completed. It will not be necessary to draw too much attention.

Victor: Still, there is a feeling that we have been noticed somewhere.

Iya: It's the hidden cameras, but I've already deleted the footage. It'll just be static on their screen.

Carsten von Munchausen-Dach: Well done. No one would even dare to say that...

Victor: Otherwise it's just an asylum.

Carsten von Münchhausen-Dach: Have you already avoided such a fate?

Iya: What are you asking about now?

Carsten von Münhausen-Dach: Were you imprisoned somewhere, after all? The authorities or the higher darkies?

Victor: I used to... But let's not talk about that now.

Carsten von Münhausen-Dach: I understand, me too, I would keep that sweet feeling of meeting myself for as long as possible...

Iya, laughing understandingly, walked forward and added:

- Did you mean with me, - and like a girl she galloped up the steps to the pulpit to the front row of the amphitheater.

Victor: Professor, what do you think: Doesn't it bore her?

Iya: The important thing is that you - Victor - are able to accept it.

Carsten von Munchausen-Dach: She's right, Victor.

Victor: This is very new...

Carsten von Münhausen-Dach: And you there, in "Africa in Siberia," are you the only one…?

Victor: This is the only way my insects read and feel me…

Professor: I see…

Iya, without waiting for the question to be voiced again, answered immediately:

- I'm fine in my woods. Wild cats come to support me when I'm lonely, and the woods are relaxing.

Carsten von Münchhausen-Dach: Do you fly over the water there too?

Iya: Yes, there are plenty of lakes. I'm in Quebec. It's massive and beautiful, and it's easy to hide.

Victor: Heard a lot about the places. Beauty! Exceptional landscapes! Spectacular nature! Just like in the old days in Siberia...

Iya: The magic of nature is enchanting... I don't write poetry, but an overabundance of oxygen often illuminates the rhyme. And now I can't resist reading or even singing a recent creation of mine to you...

Carsten von Münhausen-Dach: A pleasure to listen to. A great prelude to getting to know each other! I would very much like to say and already declare: "my friends".

Iya: So! In honor of our accidental and at the same time planned by the Highest, and instead of a glass of champagne I raise my joy and want to read or sing to you, my dear friends... Oh, how I want to call you so too and dedicate to you this "Summer Breeze"!

Victor: Let's call it our Summer Breeze! We're friends! At least, we were a couple of centuries ago.

Carsten von Münhausen-Dach: A beautiful addition, Victor! And what an accurate one!

Iya: I agree, let it be "Our Summer Breeze".

Mountain Rivers,

Summer in Quebec.

Writing stories,

Singing songs.

Jumping into the sea

Of mountainous expanses.

A fairy tale in Canada

We'll find it all here!

Huge Stones

They are all giants,

Old growth Pines,

Looking up into the sky.

Near the edge,

Freckles on his face.

With Baba in the hut.

They'll give you a pot.

Well, you don't want to.

You'll laugh from the smoke:

The tummy's churning.

Across the creeks,

Over the frogs,

Under the waterfall.

That is the sunny moment.

Someone's chirping,

It's bubbling somewhere,

It drowns out everything.

A bustling expanse!

It will,

It'll howl,

It'll flood,

Then it sprinkles.

A splash of happiness

Wash away the bad weather:

Life, Joy,

Hello!

A magical moment!

All of a sudden, it's all up in the air,

It's going to get all twisted up all of a sudden,

We're all bouncing around

Across the rocks in the water.

Let us soar

Or slip,

Squelching happily...

Our summer breeze!

Professor: You are a comedian, Iya, it turns out, but that's a great introduction to life in Canada, I hear! Thank you!!! Now it's my turn to present my mood and creation, but already at the symposium.

Victor: We'll help you today.

Carsten von Münhausen-Dach: Thank you, colleagues, and let me say, friends!

Iya: Yeah, now build a chain of invisible communication according to Claude Shannon, and everything will be easy.

Victor: Who-who?

Carsten von Münchhausen-Dach: The important thing is that sparks then do not burn out our endeavors and the information is delivered to its destination without interference and noise...

Iya: Or… just the recipient.

Professor: Yes, and without interference or noise.

Iya: Yeah, with the father of communication - Claude Shannon - we could build an amazing team right now!

Victor: So, who is he?

Everyone in the hall was already assembled. The three hurried to take their positions in the form of an invisible triangle. At the top of it was the professor. In the center on the pulpit in front of the big screen, looking around at the audience, he waited for silence.

An inscription appeared on the screen behind Carsten von Münhausen-Dach's back:

"The Divine Freebies" is a report on a flying antigravity apparatus by Doctor of Physical and Mathematical Sciences, member of the Academy of Sciences, multiple prize-winning quantum physicist...

The base of the triangle in opposite corners of the hall, at the edges of the amphitheater were held by Victor and Iya. Meeting his eyes with the professor, creating invisible threads of connection, distributing energies, calming the wave noise of the masses, straightening up and taking a deep breath, he and she together with the professor, restoring harmony with the masses, started to work.

The momentum seemed to emanate from only three points. The middle was empty, despite the apparent fullness.

The wave pulse of awareness and sense perception was 0.27. Sensors mounted on the apparatus placed on the handle of the antigravity platform, which recorded the degree of consciousness perception by the masses, ranged from 0.01 to 0.027. Not despairing, Carsten von Münhausen-Dach still continued to report on the possibility of humans flying.

Concluding the theoretical part of the report on mass production of anti-gravity vehicles "Divine Freebies" for the population, the professor asked the technical assistants to display engineering drawings of his flying platform operating on tubular cavity structures.

Suddenly, moving on to the practical part of the report and its demonstration, a pulse of awareness stunned the audience with a rattling wave sound outside the triangle, emanating from an undefined point outside the amphitheater. People involuntarily covered their ears. And, as if on signal, they covered their eyes and mouths with their palms against the bright light that flashed on the screen.

At that moment Victor and Iya quickly prepared to fly down on their platforms to the professor, who was already rising above the pulpit to meet his friends. Opening their eyes and mouths, the audience synchronously made a sound of surprise.

The sensors that record the awareness of the information perceived by the masses oscillated synchronously from 0.27 to 100 in all three on the flying machine.

- That's impossible! - Some male voice shouted from the crowd below.

- It's a hologram! - A woman shouted back.

- What kind of show is this? - A bass from someone else's voice supported the woman.

- Where are we? At the movies or at a symposium? - the chairman of the committee jumped up, offended.

Grouped together, the three landed on the stage in front of the screen in the pulpit and, closing the antigravity platforms, prepared to finish the professor's speech.

- Dear colleagues, - Iya proceeded to return the audience in the hall to a conscious state. - I support your indignation at the professor's overly unusual and excessively creative presentation.

Victor, sensing a new wave of indignation, immediately intercepted it and, defending Iya, hastened to add:

- Colleagues, I fully support what was said about the distinction between cinema and science, reality and illusion.

The professor, delighted by the enthusiasm of his support-
ers, smiling, finally hastened to conclude his speech:

- Consider, dear participants of the symposium, what you
see only as a hypothesis and no more.

The three scientists quickly said goodbye and headed for
the sea.

Without agreement, they synchronously opened their anti-
gravity platforms and rose above the raging sea area.

Seven days were swept away like seven hours. The three scientists quickly agreed to stay together and create a new school, called Vicinema Online Visual Art Academy, or VO-VAA, to represent the long-held concept of a new model of learning, still not accepted by either science or outdated educational institutions.

Carsten von Munchausen-Dach: Friends! We are alone in this world!

The only ones! We are the pioneers! - Raising his glass of champagne, Carsten von Munchausen-Dach joyfully announced to Victor and Iya.

Victor: And how divine and natural that Iya is with us!

Iya: Thank you, my faithful knights, for your labors, you will always be rewarded with a flow of mutual love under the protection of Nature!

The three, adding wind power to their platforms, swung up and swirled over the sea area in an instant.

While the three thought of a new system of education in a world society which was once again transforming itself, the ministers were unable to decide at the summits how to deal with the problem of overpopulation and control of peoples by merely reducing and exterminating their immunity. The only true solution for them was the reduction and enslavement of the biological functions of the organism of the people, in order to permanently reduce the activity of individuals and thereby officially put them on the widest possible range of drugs. In their powerful and closed circle, they decided to increase their capital by means of long-acting and long-duration biological weapons. The previous one did not give the desired results. The population was increasing, not decreasing. The only solution was – the war.

Only the three of them could clearly see what was going on around them. Hoping to dispel at least a little of the enveloping smoke of the stupefied people, the Professor, Victor and Iya continued to develop the Unified Anti-gravity Flying Apparatus, or UAFA, and the communication scheme of the new entity in the new Vicinema-VOVAA time stream.

Claude Shannon's communication scheme, developed a century ago, came to the rescue.

Claude Shannon: Friends! Whom are you here to oppose?

Iya: Turns out we're not the only ones here, Professor!
- Iya recognized her old friend as she finished her champagne with a smile.

Claude Shannon: So still, who are you here against, friends, gathered together?

Carsten von Munchausen-Dach and Victor: Now against gravity..." the professor and Victor replied, looking at Iya in surprise, taking a seat on the seashore.

They opened their mouths in surprise, looking questioningly at Iya, smiling at the emotion, overwhelming her.

Carsten von Munchausen-Dach: And you? How did you get here, dear friend?

Iya: Claude Shannon! The great Claude Shannon! Yes, esteemed genius, how did you get here?

Claude Shannon: Iya, as you can see, I am floating, just like you! - He also quickly pulled out his folding anti-gravity platform and joined the friendly company over the raging sea.

Iya: How? You too?

Victor and Carsten von Münhausen-Dah, along with Iya, surrounded Claude Shannon and had already gathered to draw him into a scientific discussion.

Claude Shannon: Easy, folks! We don't need any unnecessary emotion just yet! I'll tell you right now, I've been watching you for a long time. I'm flattered, especially by Iya's attention to my work.

Carsten von Munchausen-Dach: Iya?

Victor: How, Iya? Are you into 'communications' as well?

Claude Shannon: First of all, it's mathematics, but Iya has very gracefully turned my theory into a social theory of communication...

Carsten von Munchausen-Dach and Victor: You are full of mystery, Iya! Share your discovery, please!

Iya: I just used Professor Claude Shannon's mathematical theory of communication to prove that too much information transferred from one source to another leads to misinformation of the recipient.

Victor and Carsten von Münhausen-Dach: To misinformation?!

Iya: Or rather, to the lack of understanding of what has been received because of the amount of information that creates noise during transmission from one source to another. It's as simple as that.

Victor: As easy as flying? Or is it even easier?

Iya: Victor, I can explain very clearly on the example of communication and model of interpersonal channels, i.e. on the micro-level, and then on the example of large groups on the macro-level.

Victor, smiling, flew closer to Iya. At that moment the wind abruptly changed its direction.

All three descended abruptly onto the waves of the raging sea area.

Victor: What happened?

Iya: Perhaps something went wrong from too much emotion and we lost the antigravity wave.

Carsten von Munchausen-Dach: And your friend disappeared?

Iya: Don't worry about him. Gravity has no power over him. He has passed away long time ago. Since that time, he is in the state that he can appear anywhere and anytime.

Claude Shannon: Friends! Here I am again with you! It's not always possible to be on the same page and wave with the wind. It's been changing too sharply lately.

Victor: Let's all go back to land.

Iya: That's a great idea. It's very noisy in here already.

Carsten von Münhausen-Dach: According to Iya, there will definitely be no discourse. Do I understand the communication scheme correctly?

Iya and Claude Shannon, smiling at each other, were already flying together in the same direction with delight.

Carsten von Munchausen-Dach, winking at Victor, shouted out through the growing noise of the waves:

- And you, friend, what are you waiting for? Bring your information wave! You know the result of the big stream now, don't you?! So, divide it into metered information feeds...

Once on land, once everyone had gathered in a circle, Claude Shannon proceeded without delay to carry out his mission, disregarding the vibe of jealousy emanating from Victor.

Claude Shannon: I am only here to save your time, friends, and to make it easier for you to communicate with the outside world and to adjust the communication flows for the Vicinema-VOVAA Academy of Arts. Therefore, as people of common sense and even more, such as scientists, we will ignore all human qualities and use only the superhuman characters, habits, and functions that have already been developed.

So, my scheme.

Iya, you remember it well and can tell us, right? And I will be interested to listen to you how my theory is interpreted by you and perceived from the outside.

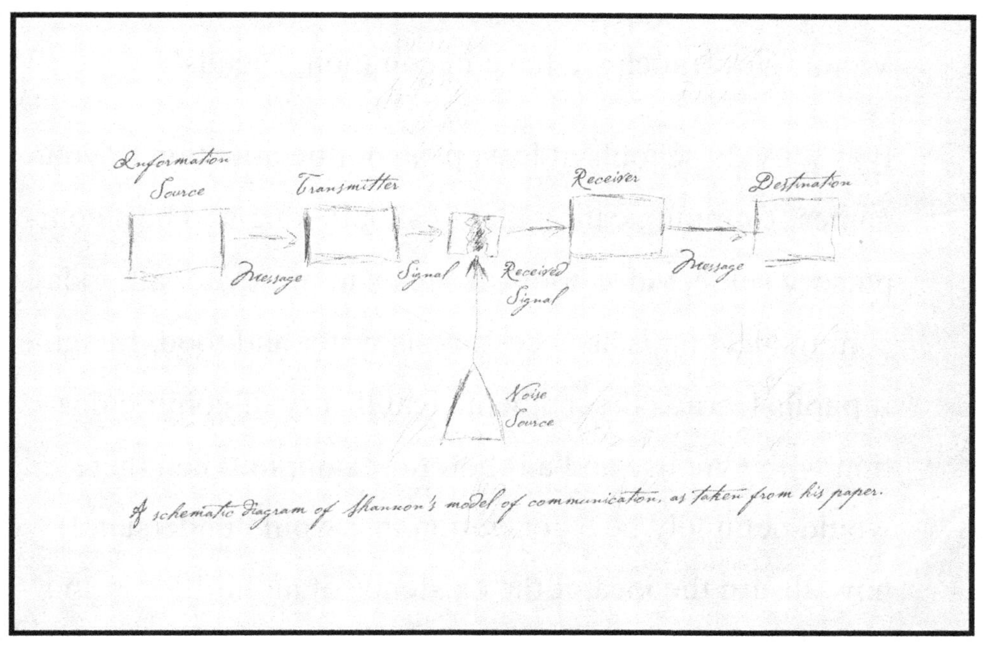

A schematic diagram of Shannon's model of communication, as taken from his paper.

Iya: I'd love to! But it's better to even fun with it and beat it a little bit, or just need to up your game.

Claude Shannon: That's my style!

Carsten von Münhausen-Dach: Oh yes, Claude! I still remember the games you created.

Claude Shannon: Really? Who was your favorite character?

Carsten von Münhausen-Dach: The heroes are the robot clown juggler and Theseus, the mouse! They reminded me very much of me in scientific circles at the time.

Victor: How much we have in common, friends!

Iya: Why do schools at least pay so little attention to your games! Communication theory should long ago be a compulsory subject in schools, it seems to me! After all, today communication is as necessary as water and food. Imagine if pupils learned the mathematical theory of communication with a mouse and a robot, for example. Then there would definitely be a generation that would understand how absurd the idea of the existence of authorities is and how the government deceives and dupes the people!

Claude Shannon: Please, don't be so bigoted, dear Iya! And yet: why do you need my mathematical scheme of communication? You already understand how stupefying the people are, don't you? Who do you want to communicate your ideas to?

Iya: We want to educate people so that people start to think a little bit about how they are being manipulated.

Victor: Wouldn't it be better to think about how to establish communication with the authorities?

Carsten von Münhausen-Dach: Iya, do you naively believe that you will want to be heard?

Victor: But somehow, I wonder if we can stop this feudal bullying of the people? How long can this go on? How many more centuries, I would ask, will this go on?

Iya: Listening to this appeal to our rulers and governors this is what came to me. The title is simple...

Take me to study with VOVAA

I'd like a modest house in Barvikha.
Or in a quiet corner, in the woods.
You could do that for three hundred million,
So as not to spin too much in plain sight.
Quiet security outside the house,
A couple of tough, smart bodyguards,
To keep the villa secure.
And taking care of my cabin.

Or maybe only: just an apartment, if there's a decent view nearby:

Schools, people, and Polyanka, where you and your friends can just live.

Or maybe here on the hill, just at Nikolenka's garden.

I'll set up my tent quietly under the tree there.

I'll have cute kids come to me with the art of living.

And in winter, we'll all be partying under the Christmas tree!

After all, we won't be asking for money from the pockets of tycoons to be happy.

We are just your servants! We want to educate you – so, let's make a peace?!

You are able to erect monuments everywhere with your power,

But we have a very simple thing here - we want to present the Academy!

For the kids and anyone else who can appreciate the game of film,

We're building an academy – teaching visual art via playing and while entertaining!

So that happily the kids will ask you to take them to VOVAA's class,

And, you don't drag them, but they beg you every day and every time:

"Take me to learn with VOVAA, I want to learn the game of film!

With mathematics and VOVAA we will make up the game again!"

May the stars rise above VOVAA!
And blesses our VOVVA.
Art and Film Academy!
Claude Shannon: A beautifully funny poem, I thought!
Then why do you need me, or rather, my communication system?

Victor: Iya knows how to make us laugh, too, as it turns out.

Carsten von Münhausen-Dach: Yes, at first, glancing at that, seems easy: All you need is a house for the academy and the support of the authorities and that's it!

Victor: And to cut to the chase, then how will this academy be different from all the others?

Iya: That's why we need a system, transmitters for effective communication: to communicate both horizontally and vertically, then also not only with the crowds or humanities but with the authorities.

Claude Shannon: You flatter me, friends! Do you think that my theory will enable you to free people from the feudal oppression of governments and create the long-awaited discourse at all levels and among all strata?

Carsten von Münhausen-Dach: I think we're taking on too much right now! Claude Shannon is right.

Iya: Our task, at the moment, is just to ensure effective communication for the three of us, for our school, and within the VOVAA-Vicinema academy.

Victor: And then we'll see what we can do.

Carsten von Münhausen-Dach: By the way, Claude, do you consider in your work the concept of effective communication in the context of the telepathic abilities of receivers, or recipients of information?

Claude Shannon: I'm sure that's exactly what Iya will be able to accomplish, isn't it?

Iya: I hope so. Especially since we have a great opportunity for empirical research, don't we, Professor?

Carsten von Munchausen-Dach: Yes, Iya! Of course, it is! But you understand that with Victor you will be able to check the purity of the experiment much faster.

Iya: That's why we need you very much at a distance, in order to confirm the presence of telepathic abilities in the absence of expressed emotional and super-emotional sensitivity to each other.

Claude Shannon: And this is another experiment for my model of communication in the context of two loving hearts.

Victor: Loving hearts?

Claude Shannon: I think I'm ahead of my time again.

Iya: It's okay, Claude. Victor is acting within the bounds of normal manhood for now.

Carsten von Munchausen-Dach: Yes, until it gets used to superhuman duties... All right, friends! We're distracted again! Now we need to get the communication signals between us to work without interference and decoders. To do this I suggest playing a game which will help explain to our pupils the model of communication of our respected friend and scientist Claude Shannon.

Iya: In the diagram, as you can see, there are three main components: the source of information, the final point of delivery of that information, and in between, the source of noise.

You got it? So, in brief: SI- SN - FPDI.

Carsten von Münchhausen-Dach: Of course, we did! But for ease of perception for an untrained audience it is better to show.

Iya switched on the hologram with a slight movement of her hand and brought up the Claude Shannon's "Communication Model" on the screen.

Iya: And now I would like to demonstrate it even more clearly with an example. Victor, would you like to participate in the experiment?

Victor: I'm willing to make any sacrifices for the sake of science.

Iya: So, I'm the source of the information and I'm sending Victor a message. Victor is the final recipient of the delivery. According to Claude Shannon's model of the father of communication, the message goes through a transmitter. In mass communication, by transmitter we mean radio or television. Today it's radio waves, Wi-Fi, and others. But we have the game at the micro level, at the interpersonal level, not at the macro level. So, friends, here's a surprise for you. Guess what will be the transmitter in our case?

Carsten von Münhausen-Dach: The question, of course, goes to the father of communication, Claude Shannon.

Claude Shannon: Iya, you have surpassed your teacher.

What could it be?

Iya: Victor, you can easily read my thoughts! Why are you speechless? Especially since you are the father of natural communication in the world of insects. Yes, Claude Shannon, I completely forgot to introduce you Victor as a scientist and explorer of Africa in Siberia.

Carsten von Münchhausen-Dach: Wait friends, not all at once!

Iya: It's a great time to introduce the game of Uninhabited Island vs. Inhabited.

Carsten von Münhausen-Dach: What is this game for and who is it for?

Iya: In this game, the plan and the concept, or rather, the teaching methodology of the Vicinema-VOVAA art academy.

Victor: Iya, can you introduce it briefly now? I'm sorry, I'll go straight to the point.

Iya: Granted. So, it's simple. The game involves two teams or groups, each with three roles: an expert minister or professor, a teacher, and a student. The goal of the inhabitants or players on the inhabited island is to create and find a prototype for an antigravity platform, overcoming the obstacles and everyday inconveniences of living in a big city and explaining the laws of physics - approving or disproving them. The goal of the inhabitants or players of a desert island is also to create or find a prototype for an antigravity platform, overcoming the obstacles and inconveniences of life in nature and in the absence of cities, and also explaining and approving the laws of physics, or again, disputing them. But also explaining the possibilities of antigravity, convincing that there is a possibility of its existence...

Claude Chenon: By the way, right now, Iya, it is important to remember the secret of my information theory: the lower the probability of events, the higher the information. That is the secret of the model and the communication system as a whole.

Victor: So, the less information Iya gives us, the higher the probability of events. Right?

Iya: Probably yes, but how does it work in our task environment? Let's take a look at it.

Professor: You know what's happening right now? Can you guess?

Iya: Yes, the information doesn't go down the pipe through the transmitter.

Claude Shannon: Right. On the other hand, information is like water. And if the vessel is smaller than the flow of information, it will start to spill out and leak out or just not all of it will go through the pipe as a channel.

Carsten von Münhausen-Dach: In our case, of course, the pipes are wide and have the ability to expand instantly depending on the flow of water or information.

Victor: Yes, we will absorb everything to the right extent.

Claude Shannon: As the super-humans with telepathy and telekinesis, perhaps, yes. But we're talking about learners and just apprentices now, aren't we? And yet, friends, go on - you have me very intrigued. So how do you explain my theory of communication in an interpersonal setting?

Iya: Ready?

Carsten von Münchhausen-Dach: Don't be stingy, colleague!

Iya: So, Victor is at the end of the communication scheme, and I am at the beginning. Therefore, Victor is the end point of information delivery. And I'm the starting point of information delivery, or just its source. And between us, according to the model, as you see on the hologram in front of you: the transmitter, the message, the signal passing through and received, and most importantly, both the source of noise and the receiver and the final point of information delivery. So, let me try to simplify. One transmitter comes from me, through which I send Victor a message. The other is on Victor's side, through which he receives the message. But between us and the transmitters there is another apparatus - producing and making noise during the communication. Could it be something or someone in our case? Something or someone that does not let bits of information through and produces too many of them or vice versa...

So, once again, as a source of information, I transmit to Victor, who in this case is the Coefficient of Performance or Efficiency, information that passes through certain, say, blocks or devices that can either distort events or something else... Victor, I have a question - a task for you: how do ants communicate? How do they transmit information qualitatively, without noise and losses? What or who is their message transmitter?

Victor: By the way, most insects communicate this way - telepathically transmitting qualitative, I would even say - vital information within their species. I mean, they arrive at the place on time and accurately. They build cities and communication systems for huge communes...

Iya: Victor, don't intrigue me, tell me, how do they communicate? At the expense of what? Or at the expense of whom?

Victor: Antennas, Iya. You already read my thoughts and knew the answer before that too. It is better not to fool us. How do you and I picture the antennas?

Iya: This is where creativity and magic come to my release.

Voila, look what kind of transmitter antenna I created!

Or rather, my nature.

Two bent spindles with wooden handles and a video of Iya working the antennas appeared on the hologram.

Carsten von Münhausen-Dach and Claude Shannon in one voice: Fascinating!

Victor: And how does it work? Video is without the sound, isn't it? Would you care to comment?

Iya: sure, for instant, I give Victor a message-question about an insect that served as a prototype for an antigravity flying platform. I direct my question to the transmitter and formulate a hypothesis: Could the May Beetle be the prototype for the antigravity platform? At this moment, Victor is in the next room or in another country and continent completely. I see that the antenna shows me: yes, the May Beetle can be a prototype for an antigravity flying platform, for example.

Victor: What if that the antenna shows doesn't match my answer?

Iya: So here we are already talking about a device that produces a noise effect or interference between the two sides of the communication.

Victor: For example, let's talk about something other than science.

Iya: About love?

Victor: We read minds, Iya. We don't need transmitters.

Iya: Exactly! We can rise above any schemes and models, which means we can soar above the machines without them!

Carsten von Munchausen-Dach: We've already noticed that. You're already soaring.

Claude Shannon: You're good! You're flying! You can soar easily! What about the rest of us?

Iya: And, for the rest we will create a VOVAA-Vicinema school where we will develop, among other things, an anti-gravity flying machine for everyone or a single anti-gravity flying machine - ELAA, so everyone will have a chance to fly. Is that a great idea?!

Carsten von Münchhausen-Dach: Amazing idea, indeed! And Claude, - our dearest friend - has to stay with us! Dear Claude, you are an engineer, a mathematician, and the creator of the future! You should help us! Now it is true that 5G does not surprise anyone. And only you, actually, also combined the supposedly unconnected things, what we are doing at VOVAA-Vicinema as well. What Iya has done now: combining science with art, and game with magic.

Iya: I would just add - the magic of love!

Claude Shannon: I love your creative adventure! I'm with you - we need to let people know how to fly!

Professor: Yes, my dear friend, Claude Shannon, it is your style to talk lightly and playfully about serious things! Ah, how nice it would be to introduce it into the new school system!

Iya: Claude! You, as the father of communication, cannot leave us! Stay! It will be fun and exciting! We have to create a new VOVAA-Vicinema together!

Claude Shannon: If you promise fun and games, I'll stay, but on one condition.

All three of them looked at the communication father in surprise.

Iya: We are in your services, dear teacher.

Claude Shannon: If your art academy is for learning and having fun, but not for distress...

Victor: For learning through play above all! Of course! That's the only way!

Claude Shannon: In that case, you'll definitely find me interested in with my mouse and my robot and the rest of my toys. Remember them, Iya!

Iya laughed like a girl.

Oh, Theseus! How not to remember! It's great that you reminded me! Great idea! That's just what we need!

Professor: I'm ready to play!

Victor: Don't drag it out, come on, tell us, friends!

The father of communication pulled out of the air, like a magician, his game invention - an electronic mouse named Theseus, and everyone with childish interest began to play, exploring the labyrinth... Then Claude, taking one look at the professor, instantly pulled out a robot clown juggler from his suitcase and gave it to the professor. The energy of the adult children's shining eyes gave them another half-century of life, returning their bodies to their youth and invigorating the spirit of invention even more.

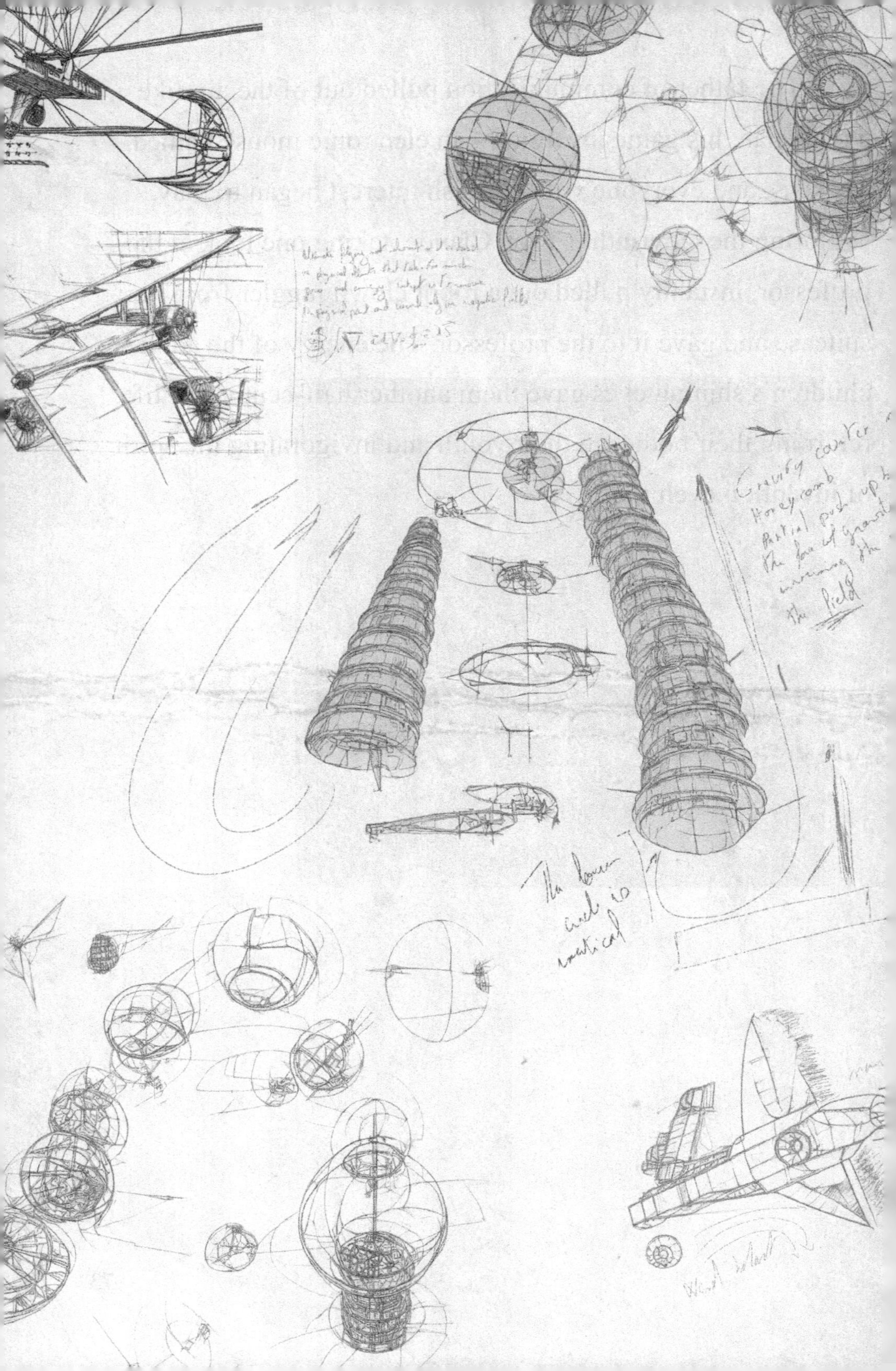

Gravity counter
Honeycomb
Partial push up to
the force of gravity
increasing the
field

The lower
circle is
vertical

Chapter two.
Victor and Iya's Spring dream.
Africa's in Siberia

Over the snow-covered Siberian valley, a blizzard swirled the small birch forest with its inhabitants. The tiny dwellings of Africa were still hiding from the Siberian cold, but they were already anticipating the rise of spring. For a second everything froze. And suddenly, with a roar, a greenish-colored beetle darted right out from under Victor's nose toward the center of Africa. For a moment the May beetle hovered, caught by the already hot flow of the African wind, but the Siberian gust nevertheless swept it even higher, spinning it in a stream of two opposite winds. In its struggle to survive, the beetle - a messenger of May, or a bronze beetle in entomological circles - diverged polar opposing winds with a resolute movement of its wings. Having slipped between the two currents, it seized the moment and, resisting the conflicting winds, landed, melted the Siberi-

an snows, outlining the shape of the African continent in Siberia...

Victor had already opened his eyes and looked at Iya, but he was still under the impression of what he had seen in his daydream. The dream did not go away for a long time yet.

After a long walk around the glade island, Victor and Iya sat down to rest under a tree in the southern part of Africa in Siberia, admiring their discovery. In the same valley- in the wood "Lesok", in Omsk - that Victor had so strived to keep from ploughing and destruction back in the sixties. In the same glade where once the May beetle and ants outlined the contours of Africa in Siberia... In the same glade of Africa in Siberia where Victor and Iya found their friends-saviors as well as inventors - both bees and May beetles, and many other mysteries of the insect world...

Iya: You said something in your sleep, smiling.

Victor: Was I asleep?

Iya: That's a good question, by the way. Where is that line between the real world and the dream world?

Victor: I see that you have put a Philosophy course and a Science of Solitude course for children and their parents

in the curriculum of our school. Isn't there too much un-
necessary information?

Iya: You think you don't need philosophy to build a flying
machine?

Victor: It depends on what philosophy.

Iya: A philosophy of wonder and nature!

Victor: Look around, how many wonders there are! Why
else would there be "loneliness"? The townspeople are
already lonely, each on his own.

Iya: That's right, they need to know how to survive on
their own. And besides, living in the city, a lot of people
are probably forgetting the starry sky, mysterious night
sounds, the freshness of the morning dew and silence...
So, what did you dream about, remember?

Victor: You know, the same repetitive dream again...

Iya: You either have an overdeveloped sense of responsibility for everything and everyone, or you really have a
mission in this life...

Victor: Mission?

Iya: Was it your dream about "the mission" again?

Victor: Do you believe in the transmigration of souls?

Iya: I even believe in the migration of continents...

Victor: You're just my mastermind! Let's call our place "Africa in Siberia"!

Iya: What a brilliant idea! That is what I thought about too!

Victor and Iya laughed, hugging each other, leaning their foreheads to each other.

Victor: That's right! And May Beetle will live there too!

Iya: Or rather, there will be a colony of May bugs.

Victor: Yes, that's right, correct. After all, May beetles always live in pairs and in families.

Iya: I noticed it too - they're more social bugs than individual bugs... But how they chew up the leaves in the raspberry tree! Just like the Colorado beetles!

Victor: You're very observant, you're the most observant! And what else did you notice?

Iya: By the way! The structure of their transparent wings, the ones under the green protective underwings or something, are similar to the structure of the leaves... Notice?

Victor enjoyed Iya's fantasies and contemplations, examining with delight the wrinkles around her eyes as she squinted to describe the May bug.

Iya: And, as for faith... or as for soul - or rather, whether I believe in their transmigration... It seems to me that when we get too bright and happy or, conversely, get angry and angry, we can certainly also connect with wandering souls who have unfinished their mission here... We travel in space, and of course, the possibility that on the way to me or on the way to you in space - in our flight - something or rather someone very close in spirit and ...

Victor: I would say someone can be caught like a virus?

Iya: Someone of a kindred spirit, I hope...

Victor: Maybe not related, but very close in spirit and

thought. It seems to me that I picked up someone familiar to me, judging by the way he describes and feels Nature.

Iya: Are you talking about the book "My World" by Victor Grebenikov that we read together?

Victor: It's so good that you remember! After all, we read it almost fifteen years ago! And it was like yesterday!

Iya: You know there is no time and space for soul mates!

Victor: How then to deny such miracles, Iya, as you and me?

Iya: This is our mission Victor. Something we have to solve through this connection. It is already good that we are enjoying this connection... There are such karmic knots that are a burden for many couples. They cannot untie them until they fulfill and accomplish their mission or task.

Victor: Then we're in luck. We don't get bogged down by our karmic connection, do we?

Iya: It's a reward for us. A gift. Isn't it?

Victor: But there's still a mission or a duty.!

Iya: The task is feasible, right?

Victor: When I was reading the book of my fellow entomologist, who claimed in his fiction-documentary "My
World" that he had created an anti-gravity platform, it
seemed to me that everything he was writing, I had already written and experienced some time ago.

Iya: All of it?

Victor: Listen to this, especially this part.

Iya: You know about the theory of a single information
space and field, where all the ideas are gathered together?
And we are just their guides...

Victor: Listen to this...

Victor took Victor Grebennikov's book out of his back-
pack under his head, opened it where there was a book-
mark made of a maple dried leaf, leaned against the tree
so that Iya could comfortably lay her head on his lap...

In this way, by reading books or studying nature, they could be together forever...

Victor began to read, and Iya periodically jumped up, repeating excitedly:

- Just like you wrote in your "May Beetle" book!

"And then a familiar vision flashed before my eyes.

"It is as if I am walking across a wide - up to the horizon - multicolored alfalfa field, dense cool green with purple, white, yellow, pink tassels of inflorescences is parted, going back, and every stem, every flower, every succulent three-lobed leaf is clearly visible. And it was as if bright-winged butterflies, big bumblebees and various bees - golden, gray, motley - flit over the field near the inflorescences, migrate from one flower to another. As I walk, I look closely at the bees on the flowers, and either pronounce or write down strange but familiar words: melitturga - one ... rophytes - one ... megachiles - two, no, even three ... anthophora - one ... This is not enough for me, I try to see, recognize among the many bees some special one, very necessary to me, but other insects flash before my eyes, greenish blue trefoils sail by, flowers go back, and in their place come more and more." ["My World". Victor Grebennikov. Science Fiction. 1997. Publisher: Sovetskaya Sibir.]

As if connecting with the hero of the book, Victor dissolved into Nature, not noticing at all how the moonlight replaced the sunlight illuminating the illustrations with images of insects and people in the book "My World". In a single flight, Victor and Iya explored their new world. There was everything that a happy man needs to be at one with Nature.

Chapter three.
Victor and Iya's Autumn Dream.
100th anniversary of VOVAA-Vicinema

Vicinema Visual Arts Academy. In the auditorium of the academy, everyone is getting ready to celebrate. In Victor's workshop his students, children and grandchildren are experimenting, creating and learning while playing.

- And suddenly - "She! " - shouting, Victor turns to his grandson, playing quietly with his net and magnifying glass at his grandfather's desk.

 - Who is "she"? - quietly interrogates his grandson.

But the grandfather, not hearing him, answers himself, picking up his picture with the bee in it:

- Oh, how perfectly fitted that proboscis is to her narrow nectar receptacle!

Turning curiously towards the picture, the grandson looks over his grandfather Victor's shoulder, scrutinizing the flower and the bee.

Victor, on the other hand, with a gasp and in his flow, not noticing at all the magnifying glass of his grandson pointed at the bee's proboscis, continues:

- I want to embrace all this beauty and lift it up... And there She is my Iya. She smiles at me and we are already flying together! Just above the commotion, like in Marc Chagall's painting "Above the City", only above my worlds and in my and Iya's flight...

- Grandpa! Grandpa! Where was I? Weren't we in Africa in Siberia hunting mysteries together?!

- A-a-a-a... You? We? You're with me, of course. You were still asleep...

- Sleeping?

- Yeah, it was very early... Fresh. And I kept covering you with the blanket. Do you remember that?

- The blanket... No, but I remember the night under the open sky well... Or rather, a few hours of sleep and you got a good night's sleep...

- Don't you remember, grandson, the summer wonders of our Africa that's we find in our wood, in Siberia? Oh! What a wonderful place to be!

- Yeah, I remember.

- And do you remember that we spent the whole day counting bees and bumblebees on that huge field-island, in our wood, in Siberia, crossing it over and over again along and marking every insect we saw with an X in the columns of our route sheets!

- It was hot in the morning and daytime then… You gave me your huge hat. I remember that very well! And the bugs in the box...

- Yes, all this under the burning sun! Then we were very tired during the day. But what a fabulous nights we had!
- Victor dreamily stretched out and again plunged into his thoughts, continuing to read passages from "My World":

"...In the middle of the night I suddenly opened my eyes: a large night butterfly fluttered its wings over my face.

Then it flew to a branch, touched the cold dewdrops over-hanging the leaves, and disappeared into the twilight with a soft 'f-r-r-r'. Somewhere nearby a belated mosquito was singing... "

Suddenly, as if waking up, Victor rushed to continue:

- The condensation makes me chilly. And I'm adjusting your blanket again and again.

- Did you stay up all night then? - Putting his hand under his chin, the grandson, having already finished his experiments with the net and the magnifying glass on his grandfather's desk, prepared to listen to Victor attentively.

- Why? I even had the strangest dream. But I had time to fly, too. Not far from where we were staying. I was still hoping that Iya would join us in the morning and I would give her our Africa.

- She never got around to exploring all of Africa? Didn't she go there once...? Yes?

- She'll make it. We'll still fly together. Only my flying machine needs to be improved. I keep thinking: maybe I should make a second seat or somehow enlarge the area to be flown? So that you and I can fly together! And with Iya I could take promenades then. What do you think?

- Did you say something about a strange dream, grandfather? - The grandson interrogated.

Victor walked over to his painting, which was of himself with a brush in hand, painting the sky and the landscape around him.

- Ah yes, a strange dream or story, and that's how it was," continued grandfather Victor, opening the sketchbook and taking out a picture to show it to his grandson.

"It's like I'm standing at a canvas, a huge panorama of the prairieland or steppe. I have oil paints on my palette and a long, long brush in my hand. I mix dark azure with whitewash and the result is blue. But the paints are stiff and unyielding, like rubber, and it's hard to mix them.

But why am I here? After all, these huge canvases, decorations, paints, paintings-all of this was a long time ago, I've been an entomologist for years-is it possible that someone has mixed everything up?

At last the blue colour is ready, just the one I need; I approach the panorama and the canvas is far away - I dab on the blue sky... But though it is painted on the canvas it is real, high and everything on this panorama, the horizon, the steppe, the grasses are all real too.

I was instructed to make the panorama better, to freshen it up, to correct it by painting it with oil paints, but the steppe and the sky are a part of the world, it is the whole world, all of nature. And there are not enough paints and they are not bright and half-dry. And suddenly I realize what a great responsibility lies on me: if I do something wrong - what if I spoil the work? What if I ruin my work? Why don't I know who and when I did *it, this work, and why I took it on? But suddenly, when I open my eyes, I see another world above me. High trees, tops already touched by the sun, a bright sky above them, not such as in a dream, but silvery, light, I see a dragonfly on the background of this sky, flying out on its first morning hunt." [Excerpt from the book "My World". Victor Grebennikov. Science fiction. 1997. Sovetskaya Sibir Publishing House]*

- Next to you, you grandson, snoozing sweetly.

"An electric train rumbled familiarly in the distance, finally bringing me back to reality and quickly extinguishing the strange excitement I had experienced in my dream." [1]

- Grandpa, you are also a romantic storyteller! - Looking at Victor's new tools with interest and choosing sticks and tubes from his grandfather's collection, said the grandson.

- It was! Don't you remember?! - leafing through his notes quickly, my grandfather tried to find Lesok's or Africa's in Siberia's sketches.

- I remember! I remember our travels in Africa very well... - The grandchild puzzledly compared the map of Africa on the wall and the map of Africa in Siberia on his

---

1   Excerpt from the book "My World. Victor Grebennikov. Science Fiction. 1997. Sovetskaya Sibir Publishing House.

grandfather's desk.

- There's no need to be sad. Let's better get ready for another trip to this fabulously hot country in our Siberia, summer is just around the corner. This time Iya should definitely be with us!

- Iya? - the grandson interrogated in surprise.

- Look how much work it is. - Victor pointed to the table with tools in the carpenter's workshop, ignoring his grandson's surprise.

- Will you show my friends again today how you make your scoop-net? My friends and Vicinema's students are supposed to come today," the grandson continued, ignoring his grandfather's constant references to Iya.

- That's great, you can teach our friends everything I have been teaching you.

- What about you? Who else could be better to teach them entomology than you? How lucky we are, after all, that the teachers of Vicinema are those who, first of all, "learned by doing" and had practiced, gained their experience via "doing", rather than "just talking" before started teaching. We are really lucky that Vicinema has amazing and honest teachers, living in their name of their scientific interests – connecting their live and practical concentrations and interests with the teaching and theory. Do you agree with me?

- May Be-etle! – The grandson wanted to say 'maybe', but some way he just managed to say "May Beetle" and then laughed at himself.

Victor noticed that and smiled back. Then, he continued holding forth and thinking out aloud:

- How else can one be honest? Only when you have walked the path yourself, and then you can easily and confidently lead your followers or scholars; otherwise, they will not believe you!

- And today?! Do I have start giving lectures right away? I didn't even get my entomologist's certificate! How am I supposed to teach my friends?

- But you have been at the apiary with me a lot of times,
you have seen and observed and practiced on your own
since you were three years old, weren't you? What's more,
you approved the Africa's in Siberia… You were practi-
cally born in Lesok! Right?

- Yes, I was. Yes, I have been…
- You've studied bees, haven't you? Have you collected
honeycombs and honey? Observed other insects? Keeping
notes and diaries? Have you mapped Africa in Siberia?
Did you map our glade-island called Africa in Siberia?
Did you study the May Beetle? Did you draw it? Did you
do all that?

- Yes, of course!

- Where are your sketchbooks? Where are your drawings?
Where are your photographs and notes? And finally, did
you make a film about the May Beetle based on my book
"The May Beetle"?

- Yes! Who else?

- Then tell me: do you have information, experience and knowledge to pass on to your Vicinema peers, friends and pupils?

- I guess, but... But will they want to listen to me? Believe and learn? Will they believe me?

- Why not? Give me one reason why you can't talk about our Lesok today? Our Africa's in Siberia?

- Well, it's just so accepted that teachers have to be adults and have papers like diplomas or at least certificates.

- Vicinema students will believe you faster than any adult with a piece of paper claiming to be a teacher.

- Really?

- I do! And you know why?

- Why?

- What do you think of?
- Because no one will tell them more honestly than I will.
- Yes! Your abided and dwelt experience is priceless. Plus, you have been reinforced by the theory and learned in practice, 'learned by doing' and then reinforced by the

Nature or the natural exam... So, meet your audience confidently and see how easy it will be for you to impart your knowledge gained in a natural atmosphere of love!

- Thank you for your trust and blessing, Grandpa! Will you back me up a little bit?

Victor looked intently at his grandson. He embraced him. He ruffled the hair on the top of his grandson's head and silently handed him the entomologist's tools: a net, a magnifying glass, two antennas for testing electromagnetic zones, an easel and a sketchbook. Then he opened his collection of insects and solemnly handed over a box with a May Beetle. Afterwards, he smiled and, leaving for the workshop, said quietly as he went:

- I need to refine the flying machine today. Iya promised to come and help me. You have been blessed by the Nature long time ago! What else could be more powerful than it?!

Then, murmuring something to himself, Victor began to lay out the drawings of an antigravity flying machine on his desk.

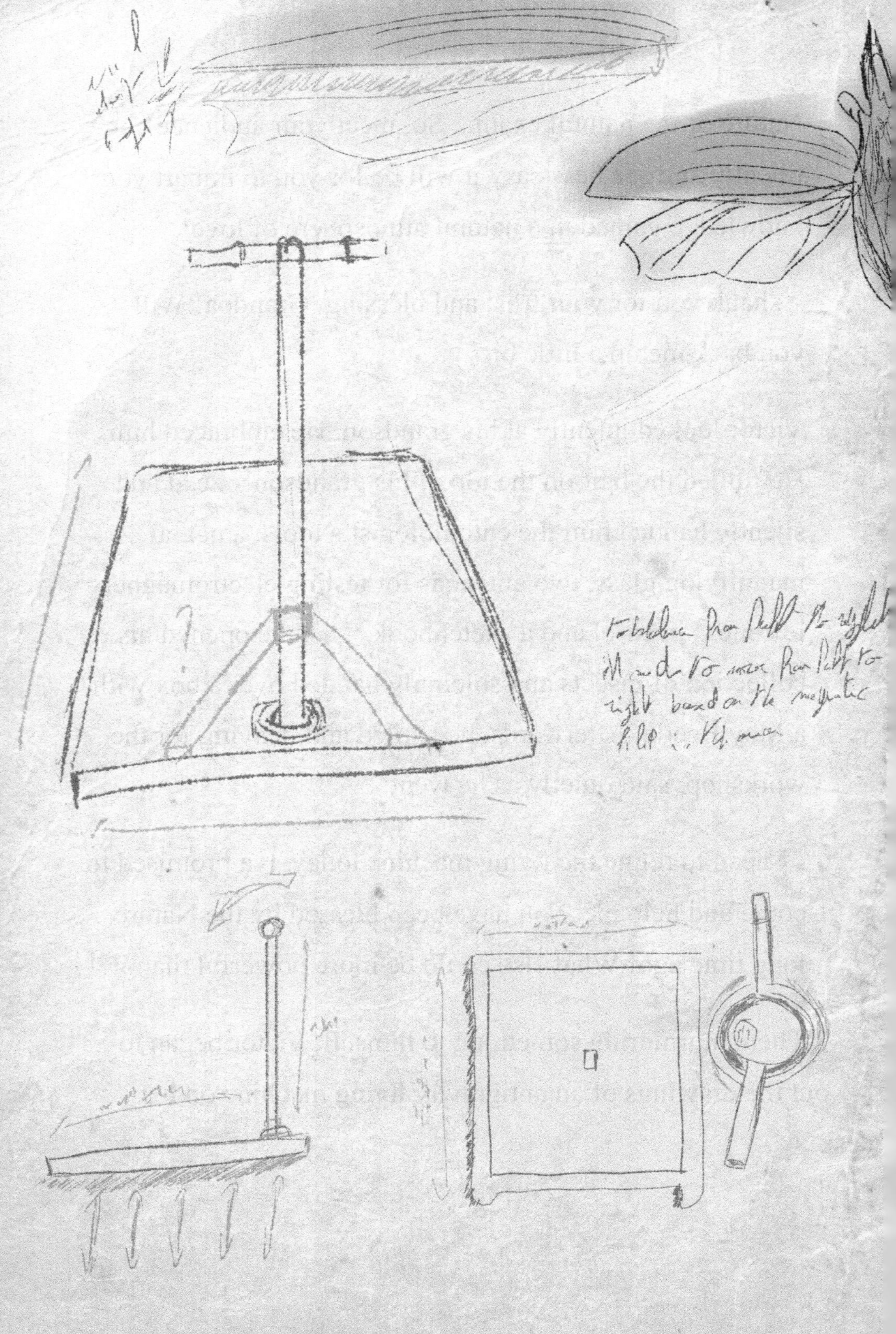
to balance from left to right
and to move from left to
right based on the magnetic
field in the same

After a short pause, the grandson continued:

- What about the lessons in levitation and telepathy, geography and entomology and all our classes today in Vicinema? - Again he asked his grandfather, being disappointed and surprised, and, without waiting for an answer to his rhetorical question, hoping that his grandfather would notice that no Iya was here, ran off into the depths of the art school workshop, mastering a new tool and analyzing the magnetic zones and areas on the layout-installation of Africa and at the same time hunting in the art workshop Vicinema for domesticated insect pets, hiding in the already mastered corners of Africa in Siberia.

- Where did you run off to? Wait!

- Anything else?

- And how will you teach your friends?

- With you! Or rather, with Iya...” the grandson joked hastily.

- With Iya? Is she here already? - Victor perked up, but seeing the expression on his grandson’s face, suddenly sat down in front of him, looked into his eyes, and hugged him by the shoulders, squinting, asked:

- Are you jealous or something?

- Whom, Iya? – Then with grin and sadness of his grandson, he gestured for Victor to sit down and have a serious talk. But at that moment Izmud and Bezmud of the staff who were busy preparing for Vicinema's centennial party came in and brought Victor's newly printed books.

- Here you are, please, take my second book, it contains not only stories and impressions of our observations but also some interesting results. There are also detailed instructions on how to make a scoop-net and feeders for ants, and everything we need for our experiments and observations of insects. - Victor hurried to give his grandson the first copy of the recently published second book "The May Beetle".

- Oh, thank you! That's great! - The grandson immediately opened the book and with an expression, solemnly began to read aloud the preface.

- Do you like the beginning?

- Yes. But I wish if all the secrets of flight are revealed

there too. It would be great indeed.

- There are clear instructions and tutorials on how to play the game "May Beetle", how to look for a prototype for the flying machine on two islands: inhabited and uninhabited.

The door ring made a buzzing sound, similar to the buzzing sound of a May bug.

- Open the door, grandson, it must be Iya! - The grandfather shouted to his grandson, screwing the control lever to the base of the flying machine.

The loud laughter of vivacious boys and girls with brushes and canvases burst into the dreamlike space of Africa frozen for a while in Siberia, that was built up and exhibited for the wintering in the house-school of Victor and Iya - Vicinema.

- You still have May Bugs here?

- Why is it still have them here? - shouted Victor from the far workshop. - They have been living here for too long...

- Hello, teacher! - The students of the Vicinema-VOVAA art school greeted him good-naturedly. And, immediately noticed, looking closely at the habitat of the families of the May beetles: - You have so not enough food for these couples. They have already eaten all the leaves of the raspberry bushes and have taken to the grapes.

- Hello, my inventors and architects of the secret and beautiful world! - Victor greeted back his students as they left the workshop. - How attentive and inquisitive you are!

- You've put up the Maybug sign again! – one of the students remarked.

- Oh, yeah, found it in the archives recently and put it up for various curious visitors out there.

- Is it just like the old dark days, with all the uninvited guests? - One of the parents asked.

- I've got an inquisitive guy named, Izmud, wondering what's hiding in my garage. Someone must have spotted us.

- And so what? Even you are noticed and revealed by them! So what? - One of the parents accompanying the group of children was irked. - Thanks God, the days of the KGB are over.

- The times have passed, but the people have remained. The call "For the Motherland!" has been replaced by the slogan "For the 'Profit'!" - Replied the fussy woman who was decorating the front hall for the holiday.

- What do they want from you now? - The parent did not stop questioning Victor.

- They can't find out the secret. They need to know everything. Billions they need to multiply! It is not enough, not enough for them! Never enough! Money is uncountable noun!! You know it is never enough! - The woman who decorated the hall and ran out to meet the children and parents did not give Victor a chance to answer, she stood with balloons and rushed into the dialogue with the parent, answering for him emotionally.

Victor, smiling, was watching the lovely women and was still in his own world, not hearing what they were talking about.

- Do you know what 'malo' means in Spanish? - Suddenly asked one of the students with glasses and a dictionary.

- And what is it? - Almost unanimously answered the question from the attentive adults in the front hall, already involved in the discussion and the pleasant bustle of the meeting and preparations for the school's 100th anniversary celebration.

- "Duro" - The cute and smart bespectacled man announced earnestly and deliberately, while looking at books in the Vicinema art school in parallelly.

- Or in English: "evil", or even worse – "grave"! – Replied nervously the goggle-eyed man's friend, following him everywhere he went.

- Let the grave billionaires dig only for themselves…

Finally, Victor started looking thoughtfully at kids' art,

paintings and the drawings, being distracted by himself. Suddenly, he looked at the guests and students, and continued in amazement: "Oh, how many of you there are today! Well done, all in one place!

- Don't you remember, teacher, you and we have a big holiday today!

- I have a party every day with my "boogers"! Come, come let me show you my latest experiment. Maybe together we'll finally solve the control problem today. Then it'll be a party for everyone! We'll be able to fly! Won't it be a celebration?!

Victor gestured for everyone to come into the great hall, leading the guests through the workshop of the art academy, where all his students had prepared a surprise for him to celebrate the 100th anniversary of the Vicinema school.

At that moment it was heard the may bug buzzing. Someone rang the doorbell.

On the move, Victor asked his grandson to open the door and added:

- Maybe Iya can still make it to the celebration.

The grandson ignored and answered nothing again. But he opened the door, greeted the guest and joined the others.

The new guest's name was Mood.

- So, - one of the parents, asked Victor, trying to destruct him from constant questions about Iya. - Did someone from the outside see you when you were flying? Did you get a glimpse somewhere? – the curious parent kept asking Victor on the way to the place where the surprise was waiting for him. It was in the big room of the art academy.

- What's so surprising? After the books were published, many people began to wonder what kind of bug helped me to make a flying machine. - Victor started talking about the beetle.

- And really, what kind of a bug is it? - With curiosity, another parents and his child picked up the theme of the conversation, already ready to write down Victor's explanation.

- That's what we're going to find out today, my friends! Isn't that why you're here today? Let's go to our main hall – the big room!

- Friends! - Began speaking Victor, who has not yet noticing that the huge space of the hall was closed by a mirror screen separating the zone of Africa in Siberia from the visitors. - To you, only to you today, I am delighted to present all my discoveries and mysteries, as well as the tools and all my sixty-year observations in the book "The May Beetle", which I have written for you! And, the beautiful Nature and my Iya will open all her secrets and mysteries, if you will manage to save her from the enemies of users and barbarians, who have already almost ruined her...

- So? After reading these books, will we be able to create many such flying machines and fly too? - Mood asked seriously and inquisitively, coming out of the crowd, and then already repeated the question sarcastically, trying to look behind the mirror screen.

Victor glanced at him, but without paying much attention to what was going on and ignoring the question, continued:

- My dear lovers and enthusiasts of Beauty and Nature, you will be able to create an even better flying machine by unlocking the mysteries of Nature. The only true key to this is your curiosity and observation skills.

- That's all, kids! - Mood sneered, already raising his voice to get everyone's attention.

Victor still ignored Mood and continued at the same mood and pace:

- And all the drawings and diagrams, of course, I will leave to you. But before I do it, I suggest we play the May Beetle game. It will also help us correct the flying mistakes. Just today my muse - Nature - will come to visit me and help me understand what is the error in the machine...

- But you have flown on it to Africa in your Siberia! - Someone shouted out from among the children in the audience.

- Yes, it is true, I tested and managed to fly on the machine, but I don't always manage to fly the way I want to. And here's another challenge - only I can fly...

Mood continued to sneer:

- Of course, he's the only one who can fly...

At that moment Victor suddenly turned around and noticed his reflection in the mirrored screen covering the surprise for the holiday and dividing the room into two parts.

He wanted to ask something, but suddenly he saw Iya.

- Here I am, your Africa!

- Iya?! How? You're here already?

- Hello, my Siberia!

- Hello, my world! - Holding out his hands in surprise, Victor approached the mirrored screen.

- Am I bigger than Africa?!

- Iya? What's wrong with you? Hello!

- Victor, really, what's wrong with me? What could be greater than our love?

- To be with you, Iya, live with you! My world is with you! This is our love!

- How beautifully you always talk, Victor!

- You're my home, Iya! The sun doesn't rise without you! Here, look at this. Are my speeches more beautiful than these perfect creatures?! Just look at them, at their wings, at their coloring, at that magically magnetic structure of fibers, patterns and lines... How harmoniously they live.

- Who? Those butterflies? One-day creatures.

- It's a saturnian butterfly!

- Ah! Those test males of Jean-Henri Fabre?

- Do you remember?! You remember that in Jean-Henri's wonderful experiments, the male saturnians butterflies flew to the female not only from the leeward side, but also from the windward side, where not even a molecule of the female's odorous substance could reach.

- Oh, yeah? How did they find her? Do you remember?

- Aren't you happy in your world without me?

- And without you, my female, we wouldn't have saved Africa in Siberia?

- You saved Africa yourself, I just....

- You are not "only", you are my world. Without you, there would be no my home, no my world, no Africa in Siberia, and most importantly, no flight!

- There'd only be your family without me!

- And so is your family, Iya.

- You know that my family is only because there was already your family. And I have always had a taboo - you are married, you have children and even grandchildren already!

- My darling, you surprised me just now! I didn't even think that...
- What? That it's not just physics that's important to me?

- But we can fly because of you! We have flown, and it's only possible because of you! You're the reason I built the flying machine!

- And we have more than a dream!

- We have a happy life - yes! We do!

- But still...

- Is something wrong? You have really surprised me today! What more could one want, Iya! Isn't it more than love? We are so happy to be together forever!

- Forever?! But are we as connected forever as you are to your family?

- Iya, my love! We're more than family! We are one! We are united! We're always in flight!

- You're flying! And I am always watching your flight.

- Not mine, Iya! This is our flight. You are my muse! You're my life and my world, my music! Together we are the universe! Two parts harmoniously joined together. We are the one! One... You are part...

- What about you?

- Me?
- Yeah, you for me?

- I... You... I told you that... I never thought about it... Wait, so you feel differently... Aren't we one? Aren't we everything, the whole world? Aren't we the one? Aren't we...

- Victor, you're so funny! Just like Faina Ranevskaya said: either everything or family!

- Of course, that's it! Iya! This is our world, our flight, this is our Africa in Siberia!

- Are we all? Are we one, Victor? The members... I mean, the family members can be one! They're the nucleus of the whole web, so they're one!

- Wait, do you want this prickly life? This marriage? Do you want to cheat yourself and everyone else and be like everyone else and walk in this circle of matter? To stay in the prison of this unit of society called family? Do you want to be this prop, this social column?

You! You convinced yourself and me that the institution of marriage is just a pillar, just one of the columns holding up the plateau with kings, presidents, and leaders sitting above us.

- Yes, I did. But I thought that our flight... Or rather, your flight we could do together. That's what would keep us together forever as creators, as pioneers, as discoverers, you know? To melt those icy icebergs of ministries, and education, and family, and politics... You know?

- It's very clear that you're obviously missing something! Really?

- Yes, there is no sense anymore! Look at the absurdity of education today, at the simulations of married couples for the sake of children and their imaginary safety, at their cowardice and laziness, at their thirst for profit... Marriage today for many is just legal prostitution and only...

- You overreacted, Iya! Why so much rage and heat! Is it only for the sake of marriage? Do you now want to destroy my world and yours by the artificially unnatural and peculiar institutions of education and family? Look at you! You said what nonsense there is around! Weren't you the one who first wanted and pursued to reform and build a new education for this new century, changing the foundations of family relationships as well? Weren't you the one who wanted to make the best of it?

Don't you want to create an arts academy anymore, where children and adults learn by playing, through films and all kinds of visual arts? Isn't that how you wanted to melt the frozen system of society, and make it Africa in Siberia?! Isn't that what you called Africa in Siberia?! Iya, we've been going for so long, we've given so much energy and love to our dream - our world – with the May Be-etle!

- You're right, Victor! Yes, the human world seems orderly at first glance, but in a reality, so chaotically silly and imperfect; while the nature and the insect world seem disorderly at first glance, but so properly organized!

- How happy I am that you understand me, my Iya!
Look at that! Look at that coupling flight of desert flies over a rock in the sunlight! What do you see?
- Two rings.

- There are two rings, but it's one! The second ring is only a shadow.

- Just like that in a marriage? Is that what you're telling me?

- And yourself, what do you think?

- You're right. You're probably right.

- Everyone knows this, but not everyone can admit it.

- What's in it"?

- For the mating circle to be in constant motion and on the fly, one of the circles or ring bearers must necessarily be only a shadow.

- You mean one in plain sight and one in the rear?

- You could say that, my Iya.

- But I am your shadow, Victor!

- You see that everything I have created, everything that truly makes me happy, exists only because of you!

Why do you reject the fact that only thanks to you I could make my childhood dream come true? And only with you I can be in the stream in defiance of all the laws of physics and matter! Contrary to all the rules and regulations of society!

- Do you still have my letter?! - Iya touched the breast pocket of Victor's snow-white shirt hiding under the dark jacket.

- How can you see everything?! You amaze me not only with your gift of telepathy... And how easy it is with you! You can read my mind. I don't have to explain anything to you... How do you do it?

- It's simple, Victor. It's all very easy and natural and simple when...

- When are the... antennas tuned? - Victor picked up Iya's words on a respire.

Suddenly someone touched Victor by the shoulder. He saw his own reflection in a mirrored screen made of foil.
- Victor, are you looking to open a surprise for you?
- Asked the organizer of the event "100 Years of VO-VAA-Vicinema", and that immediately brought Victor back to the real world.

- It seems that my surprise had already found me. - Squatting down and folding the letter from Iya into a pile, Victor, blocking the way of the ants, began to gather them into a new temporary paper shelter.

- What happened to him? - Whispering crowd of parents behind Victor's back made distracted Victor and made him focus at the same time. - We must be sure to entertain him. It's not far to the loony bin!

- Did you see that? - Quietly asked one of the parents to the other, peering curiously over Victor's head. - What is he poking around in there?

- He's picking seeds or something?

Mood happily picked up the slight wave of puzzlement from the audience, suspecting that Victor was a little off his game.

- How can a man like that call himself a scientist?! Look at his absent-mindedness and incoherence... - Began Mood loudly his provocation.

Suddenly, a new assistant of the party organizers, named Bezmood, appeared in a white cape from the crowd, and greeted Mood, who was in a shiny black suit:

- Hey, buddy, what are you doing here? Don't you have a drone test in the field today?

Everyone looked at them.
Victor, as if he was in another space, still without anyone noticing, continued:

- Look how they're greeting us! Wow! Today we are going to start our meeting, friends, with a lesson in telepathy.
- Turning to the students and parents, Victor straightened up, lifted up a bag with large black ants, and solemnly invited everyone to the lesson, without noticing the prolonged silence in the hall before the mirror screen, to open the celebratory concert and dinner in honor of the centenary of the VOVAA-Vicinema Arts Academy.

What's more, our faithful "nigers" will be our teachers today. And they have come out to meet us now, just as they did long ago, back in the sixties, before the opening of our

Africa in Siberia. Do you remember, Iya? - Victor turned to the mirror screen, turning to Iya.

Mood was about to say something, but Bezmud, taking him sharply under the arm, led him out into the corridor of the school and then out the door.

Not at all, being embarrassed by the sharp wave of whispering of guests and students, Victor headed to the center of the mirrored screen, giving his hand to Iya.

- He's a madman! - Mood shouted before the door closed.

The awkward silence and shock of the audience, the grandson quickly tried to break.

- Grandfather! -He screamed excitedly.

Victor looked questioningly at his grandson.

The organizer of the party and the grandson ran up to the mirror screen. Finally, realizing what was happening, they solemnly raised the curtain and were about to give the opening speech.

Suddenly, Victor quickly opened the flying machine, climbed onto the platform, turned the lever and soared over the layout of Africa in Siberia, shouting from the rising platform to the guests, parents and pupils who froze with delight, looking up at Victor's flight:

- Just like that, Iya and I soared over our Africa in Siberia! - Looking into the center of Africa in Siberia, Victor turned to Iya: - Iya, how happy we were, do you remember, Iya!

- Grandfather! - The grandson shouted again.

Trying to distract the audience's attention, he quickly darted to the center of the layout of Africa in Siberia and started waving to Victor with a smile.

He froze for a moment and slowly landed, looking into his grandson's face sadly, a little confused and surprised, but after a pause he resolutely handed him a bag of ants and invited everyone to observe the telepathic abilities of these insects.

After looking around the crowd once more and searching for Iya with his eyes, Victor quickly began, quoting Viktor Grebennikov's notes in his book My World:

- "Lassies fuliginous is the Latin name for another species of lasius, meaning "black as soot," as opposed to "niger," which translates simply as "black. Some entomologists call them either "fragrant ants" or "little wood ants" - we will call them here simply fuliginous. In those years, the fuliginous lived here in the Woodlands, only in the Dragon Mountains area..."

- There they are, right there. See? - Victor turned to his students and showed on the model of Africa in Siberia the habitat of this species of ants, then again quickly looked around the crowd in search of Iya. But after a short pause he continued:

- I had a long-standing friendship with ants of this species, which is described in the book "Mysteries of the Insect World". In the forty-first year I saw their track in our Africa in Siberia for the first time, and at first my heart sank: I thought they were gatherers, which lived in my childhood yard in the Crimea. When I bent down, I realized that I was mistaken, and got upset. But something brought me back to their path, and our long-term friendship with them began.

"Those fuliginous are alive even now, their nest is under the roots of an old maple tree; however, people have tried to drive them out, their family is alive and friendly and has been around for half a century...

Fuliginous is noticeably larger than its fellow ladybugs, very black and shiny, as if covered in varnish. They feed mainly on aphid syrup and their roads, leading from their nests to plants with aphids, sometimes stretch for many tens of meters. These slow ants, gleaming with tiny reflections of the sun on their tarred-black bellies and heads, prefer to walk in very narrow columns, strictly in the same place, year after year.

- And I kept them in this very dwelling," Victor turned to the guests of the academy.

- Who's so good?! Who cooked everything so perfectly? Where on earth did you find this dwelling of my fuliginous? - Victor turned to the crowd of children who had surrounded him and were watching intently as he released the ants from the bag into the new spiral dwelling.

- We made this spiral from your drawings in the book, - said one of the smart Vicinema learners proudly.

And Victor, in a friendly way ruffled the boy's hair, smiling, continued:

- It was in a dwelling, with a spiral track of thick paper,

which led to a feeder that was one meter from their dwelling, that I kept them in my childhood home in the Crimea.

Here it is! You did it very well! Exactly like on my drawings. Well done!

- And why in Crimea, but not in Siberia?

- I was born there. It was also where I discovered this mysterious world of insects, which was then confined to my backyard. But it had everything...

Victor thought dreamily for a minute, but the organizer of the party, noticing the bored parents and guests, quickly reminded everyone why everyone had gathered:

- But now we have Africa, and we can not only travel in its mysterious world of insects, but even fly over it! Isn't that right, Victor?

- Ah yes, of course! - Victor again hastily glanced around the audience in search of Iya and excitedly continued:

- Today is just the perfect occasion to introduce the new disciplines of our Vicinema Academy through play. And I hope Iya will help us in this! Don't you?! - Shouting this and smiling, Victor started nervously walking around the layout of Africa in Siberia in the concert hall. Then he stopped, sighed, stood up on the flying machine and continued, soaring like a flying carpet over Africa in Siberia in front of the guests and students of the academy:

- We will present new lessons in telepathy, levitation and telekinesis, as well as geography and astronomy, entomology and biology and, of course, physics, with understanding how our mysterious world works. And in this, of course, we will be helped by Iya and our friends - Professors with a capital letter! They are very brave people. They were not afraid to go against the tide and offer alternative education, for the first time in the history of physics they dared to doubt the truth of many of its laws. But it is not for me to tell you about it, and I pass the floor to Iya and our friend, Professor Carsten von Münchhausen-Dach.

Professor of quantum physics, Carsten von Munchausen-Dah, inclined in his antigravity apparatus into the center of the hall over Africa.

- While Iya defends the vast expanses of creation and prepares courses for you called "The Theory of Nature and Naturalness", "How to Learn to Be Alone", "Lessons in Telepathy" and many other disciplines, I offer you, my friends, to meet a Teacher with a capital letter, a scientist who dared to find errors in school textbooks for the first time in the history of physics.

- Cheers! Hurrah! - The pupils of the VOVAA-Vicinema art academy school shouted in unison. - Hooray, no more physics!

- Don't be a hurry with your conclusions, friends! - The school teacher and guest of the academy, entered the discussion at once. - We would say: no more inexact and boring physics! Now only experiments and tests. And what do you need for that?

Carsten von Münhausen-Dach instantly picked up on the discussion:

- On the contrary: a lot of physics, I would say, game and practical physics. You are absolutely right, colleague Izmud!

- And, of course, a perfect knowledge of the theory will help us all prove its imperfection, - added smiling Izmud, the school teacher.

- And we can't do it without Iya here, of course! - Victor soared on his platform, joining the discussion of the teacher and the professor soaring over the vastness of Africa in Siberia. - Iya will both lead us and reveal all the secrets! If, of course, she will be favorable to us, and we will protect and defend her as caringly...

 Victor's speech was interrupted by a slight buzz in the hall.

Everybody was expressing their confusion. Victor looked at everyone, lifting up on his antigravity flying machine, and solemnly announced:

- Now is the perfect opportunity to demonstrate how the laws of telepathy work. Today we will start our first lesson with a game. And Iya will help me to do it. - Victor said, raising his voice a little. - Iya and I will demonstrate how telepathically it is possible to transmit and read thoughts from a distance, performing tasks without any aids to communication. Friends! Today for the first time we will learn to communicate with Nature! You will learn how to communicate naturally! Just as we were born to communicate - without words. Just like a loving mother communicates with her children when they are in the womb and how many living beings around us communicate with each other. This unique discovery was made by me when Iya and I were observing ants, butterflies, bees, beetles and wasps. The mysterious world of our friends and Mother Nature reminded us of this gift given to us since childhood, if we gently and lovingly preserve and protect nature.

Victor again began to search with his eyes among the guests, parents and children Iya. Victor's grandson, feeling the awkwardness of the situation, hurried to help his grandfather again.

Everybody already unwittingly noticed the correlation between Victor's distracted attention during his search of Iya, and the increase of the gravitational field power, which led to the decrease of the flying machine. At the same time the antigravity platform rose sharply upwards when Victor concentrated on the subject of insects.

- Grandpa, why don't we start the telepathy experiments with you and me? We're related.

Victor looked at his grandson in surprise.

- Yeah, you're right. Let's give it a try. But it takes a little training, a little practice, and something else that's hard for me to explain yet, but somehow with Iya I always manage to telepathically transmit thoughts.

- When was the last time you played telepathy with Iya? – Somebody from the crowd asked Victor.

Victor went pale and let the flying machine out of his hands, but he recovered and managed to land it in an area of Africa in Siberia called the Dragon Mountains.

- Very good question! - Izmud hastened to save the situation. - But first let's ask Victor to tell us about the laws of telepathy existing in our friends from the mysterious world – ants and the other insects.

Izmud pronounced the word "koziavok" so funny and long that all the children burst into contagious laughter, which immediately defused the tension in the African-Siberian feast.

- Great idea. - Victor, agreeing, enthusiastically began to look for ants in the African-Siberian valley, aptly placed between the countries of Africa, separated by ant paths.

Those were originally enclosed by transparent tunnels, created to showcase the African glade-island in Siberia.

Victor had already begun to demonstrate his telepathic experience, when suddenly an irritated and surprised voice sounded from the crowd.

- Wait a minute, good people! - Stepped forward a businesslike older man with a beard and thick-lensed glasses. His stoop and general appearance suggested that he had spent most of his life sitting behind a desk. Under his arm was a thick folder, from which crumpled sheets peeked out and pencils fell out as he advanced to the front of Africa. - Am I the only one, as a complete newcomer here, who doesn't understand the game and the nature of what's going on, or has everyone gone crazy here? Please explain to me what is going on here and what Africa has to do with it.

Everyone went silent.
Victor approached the indignant man and, embracing him, asked:

- What do you actually want to be today? A student or a teacher?

- Me? I'm a teacher, but I'm from the Ministry of Education! Do I look like a student?

Suddenly one of the children remarked indignantly:

- What's wrong with being a learner?

Suddenly the lights went out and suddenly appeared again. The surge of electricity cheered everyone up a little and revived the atmosphere.

- My friends! - Victor smiled knowingly. - Now we are going to be mycologists. Pay attention to our little teachers, here they are waiting for us in their mysterious world. And who can guess, with what we can attract and hold the attention of our little friends? Who can guess? This is what children usually ask for?

- Candy? Candy? - I heard the thin voices of the girls already wandering around the Cape of Good Hope.

- That's right, travelers! Stay on course for the Congo territory! This is where I found the city of Lasius niger ants.

Please, all of you, come to Central Africa now! Just here is where the tall, up to half a meter high, densely shaped bumps of regular grass are located. These fourteen cities of Lasius niger ants were everywhere, including the Ruins and the shores of "Lake Chad"; two were located in "Mozambique", two in "Nigeria" and one in "Morocco". And you will study this species of ants on your own. For this purpose, I have written another book on ants, to be brought today by the publisher.

And while we wait for him, as we agreed, let's observe the ants' closest relatives, who will tell us the secret of telepathy.

I present to you, my friends, the yellow sod ants Lasius flavus - they are dispersed all over the continent, but mainly concentrated in "Libya", "Sudan" and "Mozambique". However, what they liked best was "Cameroon", which is the western glade of our "Lesok".

I invite you to wander through "Tanzania" and "Zambia". Here I usually collect for my herbarium those strawberry and briar leaves in which the megachurch bees have made clippings, and those that the May beetles have tasted.

Just then, I bent down to pick a strawberry leaf and saw a wide ribbon of red wood ants winding their way along the ground between the plants.

"Those running south - empty, but some excited, hurried; those going north - dragging a white cocoon with a chrysalis on it. They are relocating... A familiar picture, I won't disturb.

But still: why are they so restless? I had to check just in case, just in case. I walk through the woods with the ant column to the north. The cocoon-bearers are clearly in a hurry, and the shells of some cocoons are quite wrinkled. What's the matter?

We passed the area of "Kilimanjaro", perhaps the only point in the forest that was not marked by nature in any way, so we piled up some old, stumpy stumps there some time ago. Just beyond "Uganda" the forest ended, and the ants and I entered the expanse of "Kenya". The ant trail led to a familiar, already mapped home of red forest ants that had taken a fancy to the edge of one of the Ruins. I did not want to put this nest on the map at first: it was young, probably not numerous yet - there was no over ground dome of twigs, only a hole in the ground - but I marked it anyway, because the family turned out to be very "crowded".

The trail took me to an old, half-decayed stump by the "Zambezi River" trail, designated by me as the home of Formica fusca ants.

It so happens that it was the insects - friends of my child-hood - that led me into this world of the unknown, from which I, who have seen much and lived more than six decades, even now have a breath-taking and regretful feel-ing: why their most wonderful secrets insects have told me not in my youth or even in my adult years, when I had enough time, but now, in the sunset...".

Suddenly a smiling old man emerged from the crowd.
- Professor?! How? Are you here? - It was as if Victor woke up and immediately straightened up.

- My friend! - Claude Shannon spread his arms, ready for a friendly hug, - You are in fine form! Great lecture!

Ready to take part in the game already!

- Did you just fly in? Is Iya with you? - Without waiting for an answer to the first question, Victor gave out a desire to see Iya immediately.

- Are your telepathic practices working successfully? - Asked the professor in passing.

- Yes, I hope so. But Iya lately... - Victor stopped quickly, finally turning his attention to the hushed audience.

Everyone was listening to and watching the dialogue intently. The organizer of the party rushed quickly to support the conversation:

- By the way, our students, led by Victor's grandson, have even managed to compose several lines of a song to celebrate the 100th anniversary of Vicinema. While we were listening about Victor's telepathic findings in the ant colonies, this is what our students came up with.

- Learning and absorbing knowledge right on the fly! - Remarked cheerfully the father of communication.

Victor, circling already on the platform in the center of the hall, happily added:

- Learning on the fly and grabbing everything on the fly!

Come on, friends! We are looking forward to the results of the absorption of knowledge in flight! Don't be long. - Victor, landing beside the organizer of the party, his grandson and his friends, hurried the performers of the new song.

The students lined up in a semicircle, circling Africa in Siberia, and sang, smiling, in swing and blues:

Education - Vacation!

Vacation - education!

What a nice vibration:

Vicinema Education!

Communication! Inspiration!

In the homeschooling education!

Communication. Inspiration.

What a nice vibration! Inspiration!

In the homeschooling education!

Vicinema - Vacation!

Vacation – Education!

That is inspiration!

What a nice vacation:

Vicinema education!

Peace in the nations!

Peace education!

Vicinema education:

Inspiration. Communication.

Vacation - education!

What a nice vibration!

In the international education!

Vicinema inspiration!

Vicinema education!

Everyone applauded enthusiastically the students of Vicinema, who delighted the guests and teachers with a casual and lighthearted performance of an improvised song.

Victor, towering on a floating antigravity platform in the air, solemnly continued:

- My young friends and their parents! The father of communication is with us today for a reason! After all, telepathy is one of the main parts of our course on effective and natural communication. And now I ask everyone to enter the fascinating world of mysteries and riddles of Africa in Siberia. Today we will take you on a playful journey where you will learn about entomology by playing with the May beetle and learning interdisciplinary science.

Africa in Siberia will introduce you not only to the geography of the oldest continent and its inhabitants - insects - but also to telepathic practices. By communicating and playing, you will build communication schemes together with the father of communication himself - Claude Shannon. Your task today is to learn as much as possible about the insect world and uncover the mysteries of flight: find clues and patterns in the natural world - see prototypes of flying machines and the antigravity machine.

And you will be assisted not only by scientific experts, cameramen and directors, but also by doctors of physical, biological, entomological and even chemical sciences. So together today, playing together, we will not only make a movie about all sorts of small insects and bugs, but also uncover the mysteries of matter, time and space.

Is everyone ready for surprises?

- Yes! - The children shouted.

- But before you and I go to the world of the May beetle to unlock the mysteries of flight, please answer me a question: who or what else are we missing here? - Victor cast a sly glance at everyone and stopped at the two young naturalist geniuses of the Vicinema Academy of Arts.

- Means of transportation! - The nine-year-old Lev answered proudly, quickly and loudly, and had already prepared to demonstrate his inventions - a flying machine and a ship.

- What we are really missing now is our Nature and Beauty, so called by Victor - Iya! - To Victor's delight, the seven-year-old young naturalist Alex answered cheerfully.

Everyone laughed and clapped their hands together with Victor.

- And without the means of transportation we can't find even Iya, - Lev continued, assertively and seriously trying to stop the exploding stream of emotions of the adult audience.

Delighted by Lev's seriousness, everyone immediately turned their attention to him.

Smiling pleasantly at everyone, pleased to have drawn attention from his brother Alexander to himself, Lev hurried to remind everyone of the competitions "Best Vehicle in Africa in Siberia" and "Best Insect House".

Victor was also quick to support the initiative of the young naturalists.

- Thank you, Lev! And, of course, thank you, Alex, for reminding us of the important things!

Indeed, how can we be without a ship and flying machines! How can we find Iya and unlock the secrets of both our small insects and the flight?

And before we hit the road, let's choose the best means to an end!

Alexander hastened to present his flying machine, as well as a ship turning into water and rising up like a helicopter. Lev, on the other hand, presented his stable ship, describing all the advantages of rafting on it along the rivers of Africa in Siberia.

Victor held out his hands to Lev and Alexander, inviting them to take a ride on the antigravity platform. The boys happily jumped onto it.

- Meet, friends, the young scientists-inventors! - Victor hugged the boys by the shoulders, and the young geniuses were already ready to comment on the size of the space.

- Not enough room for three bogatyrs and heroes, even so small-sized as we are, - Alexander said seriously and firmly.

Lev, shifting his eyebrows, stepped off the platform and added:

- The machine is not yet ready for a three-way, Master. You still need to work to increase the space, in my opinion.

- Good thinking, colleague. I completely agree with you and your brother. I will improve the machine. - Victor firmly and respectfully shook hands with the young inventors.

Everyone applauded, and Alexander and Lev bowed to the audience, playing their roles knowingly and with a sense of humor and dignity.

The solemn atmosphere was dispelled by the teacher Izmud. He came out into the center of Africa in Siberia.

- What about me, friends!

- Is something wrong, my friend? - Professor Munchausen-Dah hugged Izmud's schoolteacher.

- I feel like I'm in some kind of abyss or interdimensional space.

- Why, you are with us and we are creating with you, all that remains is to add Iya to our company and everything will be natural! I'm sure you will feel much more comfortable once you get to know Iya and feel her vibes and attention. - Professor Munchausen-Dach hastened to reassure his colleague.

- No, it's not natural yet. - The schoolteacher continued to object.

 - Dive into Iya's world and everything will fall into place, - the professor advised with patience.

- I'm here and I don't seem to be here. And there is no desire to go back at all.

- Of course, I understand you. Once you have taken off and risen, feeling the freedom of creativity and flight, no one wants to go back to the meaningless mechanical past, moreover, full of noise.

- Oh, yeah, just think if all the cars were at least... at least electric! It would be quieter.

- Yes, silly things sound stronger in silence. They can even deafen the conscious ones. That is why they are not in a hurry to change everything quickly.

- Nobody wants to see and accept their foolishness. They even somehow, I would say, rush back, degrading. And instead of creating, they, on the contrary, started to destroy...

- You, - dear professor, and your friends, - all of you are flying, while I am only at the beginning of the way and have not even fully understood how it is possible. The only thing that it is necessary at least is to destroy illusion of firmness of axioms of physics... Or, at least to question them, to recognize that not all works so unambiguously or not all laws work in physics already today... Or, perhaps, it is the time when we can see and hear, having opened our consciousness...

- You see, you have already agreed and even realized your mistakes. And now there is very little left to do - to create a machine. We need you very much! You are that missing link, a conductor, or, to use the language of Claude Shannon, a "transmitter" that will explain to everybody that flying is possible. They think we're crazy. But they will see you.

- You speak so confidently, Professor.

- I see and I know, my friend! - The professor smiled and patted the school teacher on the shoulder. - How to translate it on your language... Well, you don't need any translation. Do you accept metaphysics and philosophy too?

- You're insulting.

- No, I'm looking for a brief explanation.

- An explanation for what?

- There! That's how easy it is! That's how easy it is to explain to schoolchildren that people can fly!

- Speak, Professor! Don't stall! How can you explain it to them?

- If there's gravity, it's easy!

- So, there's also antigravity! Through opposites? That's it?

- Yes, everyone recognizes Hegel. They would not argue with him, just as they were afraid of Newton.

- Simple genius! Gravity is there, so antigravity is there too! - Izmud scribbled something in his notebook.

- See, it's already easier! And it will be easier for you to convince them after all! You have just come out of their sandbox. They may be angry and disbelieving, but they are already squinting in this direction. They knew you were normal. - The professor looked slyly at Izmud and winked. - I hope, at least now you will not hang for it?

- All that's left is to build the anti-gravity machine before their eyes, - Izmud stated emphatically.

- Don't do it in front of them, - Carsten von Münhausen-Dah changed his tone as he ended the conversation in earnest.

- Oh yes, - grinned the schoolteacher, - or, I won't finish it.

- This is both metaphysics and esoteric, too, by the way: the lower forces develop, spread and pick up common people faster than the higher forces...

- Absolutely, right! I somehow did not look at the solution of the problem from this side.

- Not allowed...

- That's right, light is stronger than darkness, though it is slower to cover the darkness. The darkness, though it overcomes it, the beam of light...

- That's good! That's what they should talk about in school physics class, too...

- Yeah, you're goanna show...

- More strength and light. You and me together now! Why do you allow and open now the space to the lower ones?

- Right, you just need to change the plates...

- You see, you're already creating! See how simple it is! As a memento of our meeting, - the professor invited Izmud to come to the workshop to Victor, - I will give you my flying machine diagrams. And your task will be to find the difference between the three aircrafts, built in three different countries and by the three different scientists from the different fields of science. Can you guess?

- How could you not guess?! This is the best thing I could

have ever dreamed of! And how did you do that? - Izmud immediately delved into studying the three schemes.

On the table, in the Vicinema workshop, on three huge drawing tables with placards: Russia - Canada – Germany, schematics of the anti-gravity platforms of von Munchausen-Dach, Victor and Iya were laid out.

Victor walked excitedly into the room, saw Izmud already working at his desk, and immediately joined him.

- Why are you here, Victor? It's your holiday today in the first place, - Izmud is not taking his eyes off the drawings, noticed Victor with his side-eye and started a dialogue.

- I thought Iya was here. She's nowhere to be found. Everyone's here and she's gone. Either my telepathic powers had just dried up, or something was wrong somewhere.

Izmud immersed himself in circuit analysis and did not hear what Victor said.

Victor thought for a moment and immediately expressed his bewilderment:

- Izmud, here you are to everyone is just a simple physics teacher, at first glance. And you also can understand schemes of antigravity machines!

- Yes, colleague. I am an engineer like Iya, first of all. We have a wonderful foundation and school under our belt, on which we built all of VOVAA-Vicinema.

- How lucky I am to have you! And how lucky our students are. - Victor gladly admitted.

Suddenly, Izmud shrieked, jumping up and thereby rousing Victor.

- Ah, Von Dach! Oh, he is the sly one! They're the same, your schemes of the antigravitation platforms!

Victor laughed.

- Izmud, my friend! Thank you for being with us...

- You do miss Iya, don't you, Victor?

- Yes, Everyone is missing Iya today. I do not hide my sadness. It is what it is, of course, you can't hide it from the Nature. You can't get away from yourself.

- Maybe together it will be easier for everyone to understand how much we need Iya here, - Izmud tried to cheer Victor up.

- I think it's pretty clear to everyone.

- Understand what exactly?

- Perhaps we have all lost touch with the Nature. Since that, we have been always missing something, we have been always trying to fill that emptiness with something; but, we have even fallen into greater "shantytown" or just a hole...

- Boom! Eureka! That's how I'm going to explain to my students why we need to learn to fly!

Victor smiled.

- Yeah. We're going to make it! You guys go to the gymnasium, but I'll keep looking for Iya. Hope, she's safe.
- Victor looked into Izmud's eyes and stood up on the platform, preparing himself for a flight to check if Iya was safe.

- Izmud! - Victor turned to the physics teacher. - You are worrying about Iya for nothing. Aren't you?

- She's safe! I'm sure. - Calmly answered Izmud to Victor, and patted him on the shoulder. - You are worrying in vain, Victor! As long as you protect her with your care and love she will be fine. The main thing is not to overdo it.

- Yes, my friend, you are right: the ultimate goal is harmony and a point of balance: not to be in minus too much or in negative mood, and not - in the plus. Keeping yourself in between – in the point of "zero". That means is to be between the two extremes, - to be "in the golden ration". In other words, maintaining harmony of the natural order.

Victor moved the control lever to the middle point of the control scale and silently soared into flight.

Chapter four.
Structure of the
Vicinema Academy.
Off-season Dream

Structure of the Vicinema Academy

Iya: When people start flying, only then will the Vicinema education system be embodied and adopted in schools.

Professor: That's too pessimistic.

Victor: However, I don't understand why there is so much skepticism and disbelief in the success of Vicinema?

Iya: Today, convincing ministries to adopt a new system of education, learning through play, is like convincing authorities and society to fly with an anti-gravity platform.

Victor: Or just convince the authorities that they are illegally taking taxes for public education from families who are homeschooling their children.

Iya: How precise!

Claude Shannon: To convince someone of something, you have to get the communication flows in the right order.

As I understand, no one can hear it directly today. You need decoders and transmitters.

Victor: Or just don't get lost in the currents and flights and keep a strong connection always, right, Iya?

Professor Carsten von Münchhausen-Dach: Things have changed dramatically. Everything has changed dramatically.

The whole system is changing. The whole thing.

Everybody changes. Everyone.

All people change.

You might say it is banal. But, not.

We began to meet the requirements of Nature. Responding to the laws of the Universe.

I can feel it even in myself.

It's not that we have suddenly begun to conform to the laws of the universe by ending to be evil and jealous. No.

We began to feel that somehow everything has been moving towards connecting us with higher forces, with the absolute, with the cosmos, with God.

Victor: With his nature. That's what you mean, Professor.

Iya: This is it, Victor. This is it!

Victor: You are unusually enigmatic today, Professor. Is there any simpler way to say how this all relates to our new school? To the new education? To us, in general?

Iya: I think I understand what you're trying to say, Professor.

Victor: All right, Iya, interpret.

Professor: I trust your intuition completely, Iya. You bring us together so elegantly - our trio, always in harmony. Without you there wouldn't be so many harmonious relationships, I'm not afraid to say that.

Victor: And what did you come up with this time, Iya, or should I say, felt it?

Iya: Victor, jealousy and all the low energies, you know, only block your creativity and prevent you from finally completing the antigravity platform without mistakes... We would have moved on to mass production of flying platforms, long ago if not for...

Victor: What? What do you want to say, Iya?

Professor: My friends, it seems that you have not been together for a long time. The level of electricity is so high between you that not just one platform, but all three can take off on this energy.

Iya: So, about the main thing! Everything is changing! Look at the structure of the society.

Victor: What's wrong there today? As there were invaders and tyrants of people, so they have remained, only they have changed names, names and signboards. They began to lie more intricately and began to steal and enslave people. It has long been known and clear to everybody. Is there anything new?

Iya: That's exactly right, Victor. But they also taught us how to adapt to the conditions, and thus contributed to our education. Then, they came up with the Tests! So that we would stop developing the habit of thinking and only find options to dodge and guess the "right answers" to those who created the Tests and became billionaires on made up Educational Test System. This way they are also teaching and training us. And we have been adapting to their "right answers", too.

Victor: Now I've understood you. And I'll continue, they didn't stop there either.

Professor: This is getting interesting. What happened next?

Victor: Then they shut everyone's mouths to reduce un-necessary apprehensiveness, communication, and brain-power.

Iya: Exactly, they created a new system of control. But

this time they overdid it, because they themselves were scared, too... Not everything went according to plan.

Victor: Yes, there turned out to be more smart guys than sheep. Only the clever ones turned out to be shy. From cleverness they put too many blocks to themselves. Right, Iya?

Iya: And this is what happened: for the first time the enslaved people realized that they were stronger than the feudal rulers. And the middle, rejected by the feudal lords, helped the people, pushed the people to see at last clearly how long the people had been enslaved...

Professor: So how did all this... come about? How does all this relate to the new education and our school?

Iya: I explain: all the caretakers have become creators and producers. And those who were once caretakers and producers have become mere enslaver. Yet, they do produce, but nobody looks at them anymore. They are no longer trusted. They are no longer seen as power. They simply found themselves upside down. They strung themselves up by their feet and turned the whole system upside down.

And it turned out even very interesting. The people listen to the people. The people create things for the people. And when the feudal lords pressurize or threaten to show their power again, i.e. with weapons, the people quickly unite and respond. There is no longer any fear of death. The people already know that there is no death. It is just a long sleep for the soul. A pause for the body, for getting new reinforcement. And so on to infinity. That's how evolution happens!

And in all this evolution, in this process, in this flow, there is no place for the old education system anymore. It doesn't work anymore. The old education system is dying on the bones of students and their parents, captivating those who still resist changing and embracing the new one.

Victor: Beautiful speech! How does it apply to our school? How does it apply to us?

Iya: **Vicinema is the natural way of learning and the natural world, represented above all in the world of film and video. It is a world of natural play. A game with a capital letter! Only a great game teaches and inspires!**

Victor: General phrases! Be specific! Look around you! Politicians have long understood and accepted this. They are already playing with the people, but only by their own rules, so that they always win, like cunning children.

Iya: How are they playing? Already? And making videos and movies?

Victor: What do you mean "how"?

Iya: What do you need first of all to make a film or a video or at least to stage a play?

Victor: A script, or a synopsis at least, or a director's plan...

Iya: The script at least! Or a shooting plan... Actors, crew...

Victor: Iya, did you decide to teach us a course in film art and filmmaking?

Professor: Wait, Victor... You're being very hard on Iya today. Apparently, it's definitely been a long time since we've been on a flight together.

Iya: I'm ready to say the simplest and most important thing!

Professor: We've been waiting a long time!

Iya: Everything is very simple! Everything brilliant is simple! The main thing is, don't get upset or offended right away. I will be sure to show you how my theory works in practice.

Victor: So, tell me your theory!

Iya: Game! Movie! Video! Scene! Staging! Lesson! Education! Everything we give the audience for educational and formative purposes is all without a rigid script, without an unbending form, without a ministry and, most importantly, without "feudal lords". Their task is only one - control and structure!

Vicinema task is education through play, through interest, through intuition, through higher love - Absolute, if you like, through creativity and creation without repression!

Victor: This is anarchy, Iya! How are we going to control it?

Iya: Control is not about Vicinema!

Victor: I liked it, and it even helped us when we associated Vicinema or VOVAA with the main VOVAA.

Iya: Which one?

Professor: Everyone has their own favorite VOVVA.

Iya: And it will continue to be associated, only with greater reverence and respect.

Victor: What was the merit of it?

Iya: **To create the Antithesis to the VOVAS. We create learning without a script. Without a framework and a rigid form. We give a plan and a goal, a vector of the game. We describe the stages. We prescribe tasks. But we do not drive into rigid "cubes", we do not give a matrix. We offer a way out of it! The authorities are not capable of it a priori.** We send only low and dark vibrations to the authorities.

By the way, Victor, you can easily find examples of higher and lower levels of insects. Insects operates both on low vibration and on high vibration.

- Good thinking, Iya! - Victor continued, laughing skeptically. - Yes, the majority of the authorities are just the dung beetles! Horned beetles, at best!

Iya: But, as you know, you can't do without them. This is their natural destiny today! This is what they deserve today! This is their behavior today! And they should have been an Absolute - that should be the function of power today. But they are only a dung beetle! This is how we can call them today for those things that have been created by the authorities in all countries lately, in my opinion.

Victor: Well, maybe some of them are just May bugs!

Everyone laughed.
Professor: It's even a compliment to the authorities!

Victor: Okay, let's just take mercy and give them Maybug status.

Iya: Yeah, give them credit already.

Professor: That's nature's way - humane!

Mood had entered the room silently and unnoticed. The same guest who had lurked at the 100th anniversary celebration of VOVAA-Vicinema and who had seemingly been kicked out the door.

The trio, sensing him, continued their discourse without paying attention to him. Mood thought no one noticed him.

The professor turned his head slightly towards his guest and continued: Iya! Great! How shall we proceed?

Iya: If we understand our place in this system today, we are clear about to our mission, and our assignment and the plan.

Professor and Victor together: And... and...?

Iya: And it's a game and a movie without a script, but for education!

Professor: So how do we do it, dear friends? Just one small example.

Iya: OK, quick example. We make a video with our child. Most people do that today to promote a product, to get publicity and subscribers. For the mentor or presenter, there's a goal, an objective and a mission for the day - to educate the student by playing, for example, to understand why they need so many toys? In the same time, the student, for example, counts, applying multiplication or division; sorts toys out by species and type, making tables and understanding why counting programs are needed; in that same moment, the student sees the work on camera, expressing himself, and understands the wisdom of the day: the more things we have, the more energy, time and effort they take up from us. That's the hypothesis we go into the field with a day of playing and learning with Vicinema. But it is only a hypothesis. The student can either disapprove it or approve it already at the initial stage of education. We do not impose the ideology of the authorities and the controlling ministries - which are the guides of the controlling bodies. We do not engage the minds of children, we do not build the so-called legal society. We give students the right to choose, to build their world. We have a set of certainties, opinions and viewpoints. We give them the right to test how it works, using the knowledge as a tool. This system is only for the higher vibrations - it's something new; and, of course it is different from the lower vibrations.

But the important thing is to understand what knowledge is and how to have effective and quality education rather than commercial one. The Education that is based on the laws of natural communication the same as communication that insect have among their groups.

Victor: Yes, but humans are not insects.

Iya: Probably not. But a lot of people don't even deserve categories of insects...

Victor: **Yes, I agree that the insect organization system is a good example to build and model a system of organization and communication in game-based learning reality.**

Professor: Okay, general scientific phrases begin…, I see we're losing the point.

Victor: You are right! I mean, where are we going to find such specialists who can be advanced and well-trained? Who can build a game? The kind of game and reality where they need to develop themselves? This must be a conscious improvisation!

Iya: Great phrase found: 'conscious improvisation'! I would even add - consciously prepared improvisation!

Besides, we all remember: when we teach others, we learn ourselves, or rather, we are being developed and improved. In other words, we promote ourselves through helping the others.

Stepping out of the shadows, Mood, without waiting for an invitation, hurriedly asked:

- Did I overhear you talking about the dung beetle?

The three looked at Mood, and Iya answered:

- Do you have any idea what this could be a prototype antigravity platform?

- Isn't it so? - Mood looked at Iya furiously.

- What does your intuition tell you? Do you want to join the flight? – Iya looked solemnly straight into Mood's eyes, and at the same time smilingly replied cross-examinations to his questions.

Ignoring Iya, not noticing her direct question, Mood hastily and eagerly inquired:

- So, was the dung beetle the prototype for the antigravity platform after all?

Iya, smiling and exhaling, gave way to Victor.

The last one, gaining more air and straightening his back, blurted out:

- You think **if you capture an army of dung beetles, kill them, rip off their wings and glue them to the base, you'll create an anti-gravity platform and therefore be able to fly? And most importantly, be able to fly?**

- So, it's not a dung beetle. I'm sure it's a May beetle! – Mood continued guessing the prototype of the anti-gravitation platform.

 - Oh, my…! Listen to him. He can't even hear us! – Victor made a whirling move on the platform.

Mood was just about ready to go back into the shadows, but the professor saved the situation.

- My friends! Mood is our guest! He's just curious and perhaps in a bit of a hurry to satisfy his desire to fly.

Mood, overjoyed at the support, rushed to set things right away:

- I just don't want to mislead our customers. The demand for information about the prototype platform has skyrocketed and...

- That's why I don't give a clear answer. Otherwise, Iya would become another victim of the scientific progress.
- Victor nervously withdrew from the room, ending the conversation.

- I certainly wouldn't give up just like that! Look at the protection I have!

The professor, trying to shield Iya from Mood, shifted his attention to himself.

- Mood! - The professor began resolutely. - We can be on the same page and in the same team, if you, of course, agree to accept a number of conditions.

- Professor, without me you are just romantically- nuts scientists, who will only remain flying in the clouds. - Mood rounded his chest proudly, gathered more air in his rounded belly, stroked his sweating bald head and continued on exhalation: - And with me, or similar to me, you can finally spread your eternal and infinite on earth as well. - Mood, proudly concluding his speech, exhaled and added: "So you're the one who has to accept my terms, but not me who has to impose them".

- Wow! I can see that's the kind of guest you are! And what are your terms? - The professor sat down.

- Very simple, as usual. You come down to us and create your VOVAA and work with satisfaction and peace of mind. And we will create for you security and a material base - and that is all. Everything is simple, as you know.

- It's really simple and easy, Mood. But let me ask you a question: why do you personally want to fly? Do you personally want, for example, to fly to Africa with me on a platform?

- The one is in Omsk? Or, rather, what is left of it in the Novosibirsk region? Or, will your machines only make a model of Africa in Siberia in the next hall? By the way, what capacity does your platform have, for example?

The professor stood on the platform, looked at Victor and Iya. Together the three walked to their platforms and, silently rising above the floor, flew into the next room to the layout of Africa in Siberia.

- Hey! - Shouted Mood irritably. - Is that how you answer at the highest level?

- Come here, please, to us, to Africa in Siberia. - The three politely invited Mood into the hall.

- It's cold out there in Africa, - Mood muttered ironically.

- It's in Siberia, after all! That's where we found it - our Africa! - The professor continued to build up a dialogue with Mood.

He fought back as best as he could and continued not listening and following to his intuition and feelings.

- How did Victor manage to find Africa there? Or, had he

made it up? - Mood muttered to himself, seeping through the sands of the Sahara and still not noticing that he was already wandering in Africa, and all three of them were hovering under the dome of the hall, using the megaphone and sending sounds in different directions.

- You do have a bit of a sense of humor, though. We should definitely become friends! - Continued the professor into the loudspeaker. - What do you see, our distinguished guest, Mood?

Mood finally reached the edge of South Africa and turned his back on Madagascar.

- Honestly?

- Of course, please, honestly.

- I see three crazy people, who are flying over cardboard hills and mountains or rather just stands, that are painted green and yellow-brown in the shape of the hills and...

- Your perspective of the vision is clear, - the professor laughed hurriedly replied, stopping Mood's outburst of emotion.

Victor and Iya supported the professor with glee.

- This is your vision from the point you are in now. - Iya flew over to Mood and held out her hand with a smile, inviting him onto the platform.

- Where did you come from? Spying on me? - Mood continued to get annoyed.

- My mission is only to guard you, watching from above.
- Iya, smiling sweetly and answering softly, circled around Mood on the platform.

- Will you all stop whirling and spinning! It made me sick!

- You can just come up with us for a moment. - Iya kept circling around Mood, hoping to energize him with positive vibes.

- Why? Why did I ever get involved with you? - Mood was spouting his displeasure, no longer embarrassed.

- Victor! - Iya shouted, flying up to Mood. - We must be sure to arrange a second seat on the platform, must not we, Victor?

- Will you fly with me or will you sit with Victor on the platform? - Iya asked Mood seriously.

- You can't keep up. We'll have to fly, - Mood said, squatting and hunched over.

- What is the matter with you? - Flying up to Mood, Victor inquired.

- Oh, my dear guest, are you afraid of heights? - Iya guessed under the dome of the Vicinema when she suddenly looked back and saw that Mood was crouched on his side in central Africa.

"Nigers", that escaped from the paper bag left by Victor at the spiral ant house during the celebration of the 100th anniversary of VOVAA, were already inspecting it with their antennae with curiosity. And the bees, which had once saved Victor's life in Siberia and had taken root in a mock-up of Africa in Siberia near the foot of the mountain, were already rushing to Mood's aid.

Mood lazily brushed off the bee and looked guiltily at those gathered around him.

- Are the bees anti-gravitating, too? Or, are they trained, like ants? - Mood was already laughing with everyone.

- Well, that's good! - The professor sighed. - That's de-fused the atmosphere a bit.

- If only it rained now, everything would be like yesterday, - Victor added sadly, but pacifically.

Iya, slowly stepping off the platform, walked over and hugged Victor.

The professor looked at the pair curiously:

- Did I miss something from the past?

- I don't have the strength to tell right now. - Victor hurriedly fidgeted, turning on the recording disc on the flying platform. - But the whole story is here.

Iya, breathing in and out of excitement, offered to leave the hall.

- It's getting dark, - she said seriously. - You'll be able to see everything on the blue hologram screen just in time. Let's go and have a look.

- Did you make a movie? - The professor inquired of Victor.

- Or rather, my rescue bees, along with the anti-gravity platform, made an action movie.

- An action movie?

- I'd even say drama with horror elements.

The hologram opened over the already calm sea, and everyone took the first line, as if in a summer theater. On the hologram screen, Victor was seen being attacked by teenagers on the border of 'Africa in Siberia' near 'Algiers'. In the same wood field, so called 'Lesok' where Victor and Iya had once discovered 'Africa in Siberia', back then, in the 60s, in Omsk, in Novosibirsk region. Brutally beating him with chains and feet, the gangsters or just bandits suddenly began to swing and run away from the swarm of bees that had woken up in the night and thus saved Victor from the wild and vicious male and even female teenagers. Afterwards the youngsters jumped into the car, model 'Zhiguli' and the car, staggering to the music, rolled away...

Everyone was silent for another five minutes.

Then the professor spoke quietly, turning to Mood:

- Why do you think those young crooks needed to beat up Victor, Mood?

- But he was saved! Iya and his bees! Weren't they? - Mood asked incredulously, as if making excuses for himself, - what have I got to do with it?

After pausing for a moment, with a little thought, he began to ponder aloud:

- Do you think the attack was planned because Victor insisted too much on protecting 'Africa in Siberia', as you call the forest "Lesok" in Omsk?

Victor answered nothing, closed the hologram, climbed onto the platform, did a few eights over the sea surface and flew over to Mood.

- Are we flying?

- No.

- Are you afraid of heights or...?

- Both.

- So why do you personally need an anti-gravity platform?

Mood didn't answer to Victor anything and just shrugged his shoulders.

Iya smiled. She walked over to Mood and looked him in the eye and said simply:

- Let's just take care of each other.

Off-season Victor's Dream

Iya: What should we do now? How shall we live? How should I live now, Victor?

Victor: Iya, my darling... you've been with me for so many years. We've been together so long... You're so smart, and we've been watching so much together...

Iya: So, what, what do you want to tell me? Look at my world again.

Victor: Yes! I am infinitely glad to repeat: look, my dear, at this mysterious for all and so familiar and dear to us world of mysteries - this is your world - the world of Nature!

Iya: Victor! I'm talking about us! Do you understand?

Victor: Yes, Iya, I hear it! More, Iya, than you can imagine. I feel, I feel your every rustle and sigh, your every thought and...

Iya: Victor, how to live now?

Victor: That's it! Like them! You see! You told me yourself. Do you remember?

Iya: What is it?

Victor: Oh, them! Look at them! Look at them again! They are here for us, to remind us every day and every time... No, better to show us how to live! To wake us up, to sober us up, to protect us from ourselves. You showed it to me yourself, my dear, remember?

Iya: Victor! You're talking about your world, your family again! What about me?

Victor: This is our family, Iya! This is our world! It teaches us! It's the best university in the world, and the school, and even the kindergarten!

Iya: You even wrote yourself already - my world! It's yours, Victor!

Victor: No, my love! It's my world, but it's also yours, and his, and theirs, and my family's, and your family's. It's our world, Iya! Well, if you give me your charming smile today, then let it be Victor and Iya's world! Would you like that?

Iya: You want to know what I want?

Victor: Always, my joy! Tell me what I have not seen, what I have missed...

Iya: Victor... we have so little time left... Do you understand me? Maybe we can still solve our main task today and find this mistake...

Victor: Iya, you're right, as always! We need to understand the main thing as soon as possible... I can't understand why the level of the machine jams when the air temperature changes and the wind is increased?

Iya: You again...

And, sighing, she turned away.

Victor: Is something wrong? Do you have to go already?

Iya: Victor, maybe you really missed something in your world?

Victor: In our, in our world, Iya!

Iya: You need to review all your notes and drawings.

Victor: With you, my darling! Only with you together!

Iya: Victor, I'm serious, here's another look at the life of cicadas.

Victor: Why cicadas?

Iya: The cicada, or rather that part of its life when it's sitting underground, reminds me of me. I sometimes think I even look like her, here, look at me. Just don't look in the eyes, now, please...

Victor: Why is that? What's wrong with you today? You don't look like yourself? And exactly what part of you am I allowed to look at today?

Iya: Only for the part of their life when they're sitting underground. My whole life with you seems to me to be sitting underground.

Victor: Why are you like that! You know yourself... If there were no you, there would be no flying machine!

Iya: Okay, then at least tell me, which of your insects do you associate me with?

Victor: Iya, I don't recognize you at all today! Look at me, look into my eyes! Just please don't leave so early today! Do you want... Do you want me to make a two-seater platform! You want a two-seater, don't you?

Iya replied nothing.

And the silence that followed.

Silence fills the Victor heart at last. However, the idea of creating two-seater anti-gravitation platform swirled his mind and kept him awaking again.

* * *

It was getting stuffy, Victor got up from the couch, walked around the drawing table and walked over to the garage window, opened the window to get more light, and lifted the anti-gravity machine up, looking from below and to the side at how to disconnect the bottom of the platform to add another part.

- Grandpa! Grandpa! That's a great idea! Hooray! We'll fly together! You'll make another place, won't you?

- And how do you know I decided to add a second seat?

- And you just finished in your drawings and said now... – Grandson replied being puzzled.

Victor rubbed his eyes, stretched. He bent down and touched his big toes with his hands. Then he jumped up and clapped his hands over his head. Then walked over to his grandson, picked him up. Looked at him and exclaimed:

- That's how you've grown! I've only just noticed...

The grandson, smiling, replied:

- I noticed that too...

Victor concentrated again and began to reason out loud:

- The only thing we need to understand now is why our machine loses control when the wind and air temperature change. Shall we solve this problem with you?

- Let's make up our minds! - With a glow of happiness, the grandson resolutely began to consider the details of the apparatus and drawings.

Victor looked around, searching with his eyes for the physics textbooks left by his school teacher at the VO-VAA-Vicinema centennial celebration. Then he saw a neat stack of books above the bee honeycombs. Quickly found the marked pages with descriptions of experiments and handed the textbooks to his grandson.

- Today Iya is a bit tired and went probably to the workshop to draw. Your task is to independently describe possible errors in the physical experiments presented by our esteemed school teacher at the celebration. And, most importantly, write down the hypotheses that break down the axioms of the laws of physics put forward by the teacher during the demonstration of these experiments. Got it? Remember?

- No. I don't remember much, - the grandson muttered to himself sadly, sitting down on the platform with his physics books and leaning sadly on the antigravity platform's control knob.

At that moment the platform went up slightly and shifted to the right. The grandson jumped to his feet in surprise and let go of his school books.

- Aren't you going to forget this? - Smiling and also wondering, shouted Victor from the far corner.

- Miracles are not forgotten, Grandpa!

- That's good. It's even good that you don't remember your high school teacher's performance in detail. Freshness of mind will help you, maybe, to solve my problem during physical school experiments too.

- Gladly! Thank you, Grandpa! I won't let you down!

Suddenly Victor felt something tickling him near his heart. An ant that had crawled out of the inner pocket of his jacket was determinedly heading for the center of his stomach, rustling its paws over the paper pile in Victor's pocket. The ant deftly crossed the borders with islands of mother-of-pearl buttons on the Nature's guardian's white shirt. The black lasius, overcoming all the obstacles in its path, rising and falling now and then along the tangled gray hairs on Victor's chest, had already passed the solar plexus of the protagonist.

Victor, chuckling involuntarily, reflexively shook off his shirt, pushing back the sides of his jacket. For a second everything disappeared from view, Victor zigzagged slightly and was a about to fail. Falling backwards was successfully stopped by the wall of the garage. Sliding down, Victor landed on something not very hard. Something cracked under Victor's weight, he recovered slightly. Groping for the object that cushioned his fall, he recognized cardboard egg cartons glued together in the shape of bee honeycombs. Like a boy, Victor, stretching slightly forward, jumped up, snapping his fingers joyfully in the acoustic space of the garage.

As he unfolded the roll and folded it into the shape of an airplane, he kissed the tip of the nose of the new paper flying machine tenderly and at the same time joyfully, gleefully. With the movement of a boy launching an airplane, he inserted Iya's letter into the breast pocket of his jacket, correcting the ends of the wings as gentlemen do with their handkerchiefs in the pockets of elegant jackets or 'troikas'.

Then, patting his pocket with Iya's flying letter, Victor headed resolutely to the blueprints for the antigravity machine. Quickly he straightened out the rolls of old blueprints. He took out some blank paper and, copying the old drawings, in an hour he was ready to construct a new two-seater flying antigravity machine with a connecting device for a second, more elegant and feminine, but the same flying antigravity machine.

By morning the work was completed. It remained to write

the name of the two combined machines, capable of connecting and disconnecting at willpower.

With a light stroke on the connecting part of the two-antigravity aircraft under the stand, the creator wrote: "Victor-Iya-Victoria".

A melody sounded, and it was echoed by the hum of a waking May beetle in the window space. The rhyme formed and splashed out in Victor's purr. The grandson picked up his grandfather's tunes in a whisper, and the song formed:

Victor-Iya-Victoria,

May Story:

School years.

To them nature is one!

Victor-Iya-Victoria,

The school's theory

It's nature's gift to us.

Their nature is one!

Victor-Iya-Victoria,

Rising to the occasion with theory:

Physics and history, -

You're always flying,

Your flight is forever!

Victor-Iya-Victoria,

Practicing the theory,

And the summers are over:

On a day in May - miracles!

May Day is a miracle!

May Be-, May Bee is a star!

Victor-Iya-Victoria!

The nature is one!

Victor-Iya-Victoria!

Their nature is the same!

And fly all summer long!

And nature always!

Victor-Iya-Victoria!

Defeating history,

Physics and theory:

The May bug is a miracle.

The May bug is a wonder!

At the sound of drum rhythms, trumpet, violin and piano all the students of Vicinema school came running, they picked up the words and the melody of the song.

Victor copied the lyrics of the song and handed them out to everyone entering the garage.

A random sheet of paper, forgotten by someone, flew out from under the printer's cover. The wind picked it up and carried it to a glade island at the foot of Africa in Siberia, where the May beetle awakened by the music followed. The bees, waking up from a long winter's sleep, exploring the discovery of the May beetle, began to diligently study the letter of Iya with their antennas, periodically shuddering and creating a light electric discharge: buz- buz- buz- buz- zzzzzzzzzz...

Afterword
Iya's Letter and Flight

When the game cannot be played...
When only the sun smiles,
To your true feelings...

"If I can't tell you, you can still one day read about what I promised to keep quiet and not spill, playing with the wind in the scorching rays of the African-Siberian sun.

Most people today say: there is no love. There is only the desire to hold on to the comfort and rear of home and family, especially now that the entire intelligentsia has long understood that this is the Third World War.

But why is it so unbearable to want to embrace and love you again? To create only with you and to tell all my ideas only to you? You repeat over and over again: "I have thought of everything". You run away from me and repeat again: "I have a lot of work, a wife and children... just like you have a family"...

There's a crowd of men around again, eager to woo. These are the right people: agronomists, farmers, gardeners, engineers, professors, Nobel laureates, politicians, company presidents, directors of museums and nature reserves, publishers, physicists, astrophysicists, chemists, artists, teachers, ministers and many more - they are so important now, to carry out my long-cherished plans for our flight and the creation of a new art academy.

But you're the only one here!

What else is also terrible is that I am even afraid to write and confess how much I have dissolved in desire to be with you... Church in former times would burn me on the stake already for only unworthy behavior as wife and mother... I understand how terrible my actions are now, I fight hard to not love you, but to love that which is given, and those who are near and perfect compliant and agreeable as never. Doctors and physicians, who specialize in female psychology and study especially important organ - brain, affirm that it is only hormonal jumps in the period when I want you as a male, and I, as a female, desire only one thing - to satisfy my inferior nature and to give birth as more little females and males as possible. But again, why do I want to fly only with you? Why does the thought of you draw me so high? To be in flight, just seeing you?!

Just a few years ago, everyone stopped noticing me and reckoning with me, envying my every little success, jealously stopping me from advancing on my path and my work, my calling... Though maybe it's not a mission? Maybe it's all just so you can feel just a little bit of how I might feel about you... So that you can just save another one of my Woods, that is so called by you as just "Lesok", a cozy little corner of my soul, a part of me... That you once again transformed that forest, that island or that glade which had been so frozen once, abandoned and already watched over by agronomists and hunters for fresh and untouched lands... But you made a miracle - you not only saved one little Siberian forest, but even turned it into the Africa of wonders. How did you manage to perform this miracle!

I do not even put a question mark, because I understand only one thing, that it is not fated to desire a man so much on this amazingly mysterious and generous planet, which with all its beauty and world of wonders indicates and tells us: look at me, at Nature - no one will ever love you like I do - your Iya, Victor!

The only thing you have to do is just be with me and keep me, just watch me, feel me as tenderly as you once did, then, by the big birch tree, so masterfully and naturally, as no one else in the world could touch the sky with a brush, playing with colors on a canvas, so elegantly and masterfully painted the only picture in your world, as unique as you...

Now I only apologize for making you play so quickly, in plain sight, in a chilly spot on the stage, without even asking you which play you would have preferred, and without asking you about your repertoire. But you understand: people have been queuing up for days, weeks and even years at the Bolshoi Theatre our Vicinema Academy if I may to compare it, and you were the only one chosen to perform on that stage that had been waiting for you for so many years... We couldn't let our chance slip away! You don't regret it, do you? That you improvised so masterfully? Well, I hope I've cheered you up a bit... At least for now...

The only strange thing is that as soon as I realized that I would like to live with you, and took decisive steps towards what I forbade myself for so long, all of a sudden as if they came to their senses and wanted to hear me... But I already allowed myself to come to you and look into your madly deep alluring eyes, which only once showed me that you are just as scary. It is just as scary to close your eyes together with me and find yourself in a completely different and fairy-tale world of ours - the world of nature and colors that you can masterfully improvise, making my world even richer and brighter, keeping it in your heart, loving it as strongly and mutually as your Nature!

Upper-Power - Absolute - God knows, how happy I am with you! My only prayer right now is for you to feel just a little bit of the happiness that overwhelms me... Nature falls in love with the Human! Hope, the Human will fall in love with me – the Nature, one day…!

I asked Nicholas the Wonderworker, and he allowed the bliss of happiness to come into my life through you! All my life I have waited only for you... and - could anyone wish for a greater reward for humility and patience in this life!

A low bow to your living mother for the beauty that she created together with your father, for the nobility and intelligence, masculine strength and dignity ... All this is so harmoniously combined in you.

May our beloved picture that we painted together, then, in our Africa in Siberia, remains with us as a memory of love for the one mine - the one that awakens me to hear and touch the most beautiful thing in the world - the music of life - together with me - natural and unified!

One

...The music of nature crowns life!

To deny Nature is a sinful thing.

Just for a few moments.

We in Flight will be destined... "

Victor followed the May bug to Africa in Siberia, arriving on the platform for the letter. He slowly picked up the scribbled sheet, reread it once more and put the letter back into his left jacket pocket. He returned to the antigravity platform, turned the control and a lever and went flying.

The time was calculated in seconds. There could be no delay. Otherwise, it would be a loss. Late. Which meant more hard feelings.

- After all, she's so vulnerable right now, - Victor said aloud. - We must hurry! We must hurry! - He repeated to himself as a mantra and continued to speak to himself: "Our Nature is so defenseless and wounded! Who, if not us, can save it! Where has the professor disappeared to? Has he stopped practicing telekinesis? Could he have ignored our telepathic signals?

Victor looked around impatiently, hovering over the vastness of Africa in Siberia.

- I hope no one's noticed me yet.

While talking to himself and trying to figure out the reason for Professor Carsten von Münhausen-Dach's disappearance, he did not notice how he suddenly found himself standing by the longest bridge in the world - the Bridge of Federations and All Nations.

Suddenly a lowering head, all covered with long dangling hair, peeked out of the shroud before him. It was the head of a girl whose eyes looked down in confusion. In an instant Victor flew up to her. There was no time to think. Her arms had already broken away from the outer railing of the bridge and spread out to fly over the ocean. Victor, letting go of the control lever for a moment, set his platform up so that he could pick up her already falling body, and under the double weight the machine flew down into the abyss, toward the ocean.

- Goodbye, my love, goodbye, Iya! - Victor cried out in despair, preparing to let go of everything and surrender to flight in the hope of teleporting to Iya's space.

Hearing the word "beloved", the girl suddenly hugged Victor tightly by the neck, and he automatically took her in his arms in flight. And they were already falling together with the speed of a stone into the ocean under the Bridge of Federations and All Nations, looking at each other with wide-open eyes in bewilderment.

- Who are you?! - Shouted the shocked girl.

- Why don't we get to know to each other? We've got time! - Victor managed to joke in flight.

They closed their eyes, and a meter into the water, and suddenly some force carried them both upward in a stream of blinding red-green light. The wave of light and wind hit Victor, knocking the girl out of his arms and somehow bumped him and Iya together instead.

- Iya? Iya, honey!

- Victor?! How is that possible?

Carsten von Munchausen-Dach, smiling slyly and satisfactory in a glory, was already climbing up, leaving the two in the same space created just for them.

- Bzzzzz, bzzzzz..., - someone muttered next to Victor and Iya.

- What's that? Professor?

- Get on his back, quick as you can, - the professor shouted as he flew away.

- It's a May Be-etle!

Spreading their arms, they flew upward.

And from above they could see the boat with the rescued girl and her captain floating below them. From the boat came the sounds of the same melody "Wings" that Victor and Iya were already purring to themselves as they flew.

The End

01.10.2021

MAY BEETLE OR AFRICA'S IN SIBERIA
By Victoria Korchikova-Malovichko

Annotation:

May Beetle or Africa's in Siberia is a sci-fi novel about love and communication between the Nature and the Human. It is a sequence of five dreams about how the Nature falls in love into the Human who observes and takes care of the Nature. Besides, it is also a model of scholastic institution – Vicinema Visual Art Academy and/or Vicinema Online Visual Art Academy – VOVAA.

Brief Outline:

A Canadian scientist Iya, the principal heroine of the story "May Beetle or Africa's in Siberia", once united with Nature, suddenly realizes that she can fly. Five dreams of a new system of education, Iya sees when she meets her colleagues at the scientific Symposium "Anti the Laws of Physics" during a presentation "Divine Freebie" by a German Professor. They create an anti-gravity machine and a new education system for Vicinema Academy or VOVAA (Vicinema Online Visual Art Academy).

Synopsis:

A scientist Iya, the principal heroine of the story "May Beetle or Africa's in Siberia ", once united with Nature, suddenly realizes that she can fly. Five dreams of a new system of education, Iya sees when she meets her colleagues at the scientific Symposium. They create an anti-gravity machine and a new education system for Vicinema Academy or VOVAA.

"May Beetle or Africa's in Siberia" has also surrealistic, comedian and romantic elements in the sci-fi novel.

Three scientists meet at scientific symposium "Anti the Laws of Physics", where one of the professors presents his report "Divine Freebie" on an antigravitation flying machine or platform. They are fascinated by the identity of their flying machines they had created, but not surprised by the coincidence. After all they possess telepathic abilities. What do the scientist agree on to make up and what happens in their flights? Where is Africa in Siberia? And, why does the main lead – Iya – choose the May Beetle? Everything is presented in dreams: winter, summer, spring, autumn and off-season.

About the author:

Victoriya Korchikova-Malovichko (pseudonyms: Viva Vichka and Rocky Fokichrock), president of Vicinema or VO-VAA (Vicinema Online Visual Art Academy); sociologist, film-maker, teacher and journalist, author of articles on education, urban sociology, and sociology of mass communication; author of prose and poesy both for children and adults: "Barefoot Across the Ocean: "Niktoshka", "How Much is Mutual", "Ya Lava Yu"; "Verse by Verse", "Seven by Seven", "May Beetle or Africa's in Siberia", and others.

Dreaming of a United City of Peace and the Sun, and having prepared a dissertation on the integrated information space, constructed by the effective flows of information and communication, the author of the book "May Beetle or Africa's in Siberia" Victor-Iya presents, in her surrealistic sci-fi novel, the main character Iya as a prototype of Nature and Education, a symbol of union of the countries: Canada – Russia -Ukraine and Germany; in the incompatible up-to-date world reality. By this, she expresses her belief that one day peace will be restored in the world, and countries and peoples will be united again in Harmony and Love, when they realize the need to merge with Nature.

MAY BEETLE OR AFRICA'S IN SIBERIA

ISBN: 978-1-9992707-9-7

Victoria Korchikova-Malovichko, Viva Vichka & Rocky Fokichrock.

May Beetle or Africa's in Siberia / Victoria Korchikova-Malovichko, Viva Vichka & Rocky Fokichrock. - Canada. Quebec: Bipolcom Inc., 2022. / Canada. Quebec. St-Jerome, J7Y1B1.

In the science fiction novel "May Beetle, or Africa's in Siberia" all names and events are fictitious, any coincidences are accidental. One of the tasks of a science fiction novel with surrealistic moments is to present in an artistic form the documentary events, inventions and discoveries made by scientists Victor Grebennikov in the field of physics and entomology, and Claude Shannon - the father of communication - in the field of mathematics and information technology.

For educational and entertainment purposes, the science fiction story "May Beetle or Africa's in Siberia" uses quotes and excerpts from the book "My World" by entomologist, scientist and artist Viktor Grebennikov. Viktor Grebennikov. Science Fiction. 1997. Sovetskaya Sibir" Publishing House. The work also uses references to games created by the scientist and mathematician, Claude Shannon, for educational and entertainment purposes.

Designer, layout and photographer: Riad Chebli

Publishing company: Bipolcom Inc. - bipolcom.com

For art academy: Vicinema - vicinema.com

ISBN: 978-1-7388835-2-3

Made in the USA
Monee, IL
07 July 2026

56551517R00115